# This book belongs to

_____
- - - - - - - - - - - - - - - - - - - - - - - -
_____

Grade: _____

School: _____

A few minutes of math workout every day will help the children master the math skills.

This **"100+ Days of Timed Tests: Triple Digit Addition & Subtraction"** is the beginner's level math practice workbook for Grade 2 to 3.

This book is specifically designed for numbers up to 999 (three-digits). This set of math practice worksheets is designed to test addition and subtraction **with and without regrouping**. The kids can challenge themselves with the timed test problems. This book mainly focuses on improving addition & subtraction and building confidence levels.

This book also has **Addition and Subtraction Answer Key sheets** at the end of the book, so that you can easily check with the kid's answer.

In this book, there are 30 problems to be solved daily and a total of 102 pages of Timed test practice sheets. It helps the kids to perform consistently and trained to be excellent in addition & subtraction.

Please check out other books **without regrouping**.

## Table of Contents:

1. 51 Pages of Timed Test Addition Sheets

2. 51 Pages of Timed Test Subtraction Sheets

3. Addition & Subtraction Answer Key Sheets

4. Certificate of excellence

## Problems Examples:

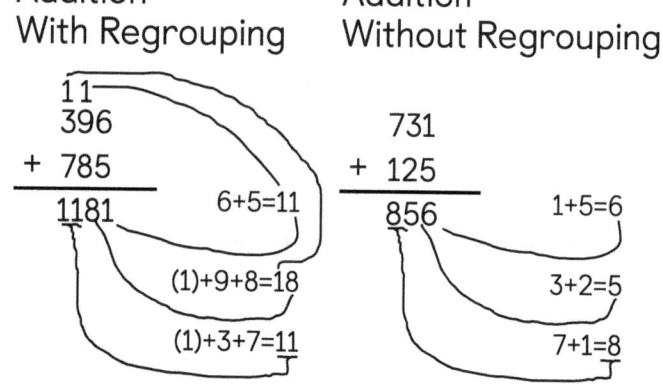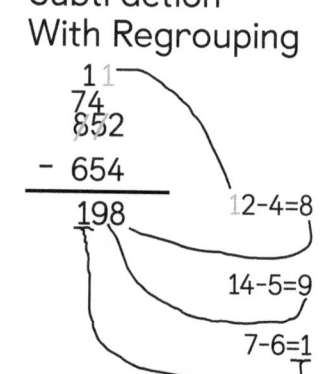

Addition With Regrouping

$$\begin{array}{r} \overset{11}{\phantom{0}} \\ 396 \\ +\ 785 \\ \hline 1181 \end{array}$$

6+5=11
(1)+9+8=18
(1)+3+7=11

Addition Without Regrouping

$$\begin{array}{r} 731 \\ +\ 125 \\ \hline 856 \end{array}$$

1+5=6
3+2=5
7+1=8

Subtraction With Regrouping

$$\begin{array}{r} \overset{11}{\overset{74}{8}}52 \\ -\ 654 \\ \hline 198 \end{array}$$

12-4=8
14-5=9
7-6=1

Subtraction Without Regrouping

$$\begin{array}{r} 952 \\ -\ 531 \\ \hline 421 \end{array}$$

2-1=1
5-3=2
9-5=4

3000+ Practice Problems

# 100+ Days of Timed Tests

## Triple Digit Addition & Subtraction

Math Excel

Ages 7-9

110 Pages

Grade 2-3

MATH DRILLS WITH & WITHOUT REGROUPING

Dear Parents,

Thank you for your purchase!

I sincerely hope, this book will be more helpful and interesting for the kids to learn **Triple Digit Addition & Subtraction**.

Your opinion matters to us. We'd love to hear from you! You can leave your valuable comments and **honest feedback.** Please take a moment to write a review. We appreciate your support.

Email us at abczbook@gmail.com with the title "**Triple Digit Addition & Subtraction**" and get your free practice worksheets!

Hopefully, your kids will enjoy this book.

Enjoy Learning!

**abcZbook Press**

abcZbook Press

A
| 396<br>+ 785 | 981<br>+ 615 | 601<br>+ 239 | 795<br>+ 757 | 647<br>+ 771 |

B
| 860<br>+ 145 | 561<br>+ 537 | 825<br>+ 830 | 213<br>+ 509 | 977<br>+ 463 |

C
| 797<br>+ 319 | 337<br>+ 441 | 419<br>+ 324 | 741<br>+ 225 | 215<br>+ 301 |

D
| 166<br>+ 435 | 622<br>+ 973 | 168<br>+ 637 | 582<br>+ 268 | 587<br>+ 318 |

E
| 765<br>+ 408 | 475<br>+ 862 | 443<br>+ 825 | 684<br>+ 422 | 830<br>+ 351 |

F
| 785<br>+ 847 | 891<br>+ 546 | 760<br>+ 392 | 951<br>+ 428 | 735<br>+ 734 |

A
$$
\begin{array}{r} 241 \\ +\ 111 \\ \hline \end{array}
\qquad
\begin{array}{r} 253 \\ +\ 936 \\ \hline \end{array}
\qquad
\begin{array}{r} 820 \\ +\ 912 \\ \hline \end{array}
\qquad
\begin{array}{r} 303 \\ +\ 829 \\ \hline \end{array}
\qquad
\begin{array}{r} 224 \\ +\ 501 \\ \hline \end{array}
$$

B
$$
\begin{array}{r} 633 \\ +\ 430 \\ \hline \end{array}
\qquad
\begin{array}{r} 804 \\ +\ 215 \\ \hline \end{array}
\qquad
\begin{array}{r} 976 \\ +\ 996 \\ \hline \end{array}
\qquad
\begin{array}{r} 846 \\ +\ 941 \\ \hline \end{array}
\qquad
\begin{array}{r} 819 \\ +\ 109 \\ \hline \end{array}
$$

C
$$
\begin{array}{r} 634 \\ +\ 264 \\ \hline \end{array}
\qquad
\begin{array}{r} 756 \\ +\ 845 \\ \hline \end{array}
\qquad
\begin{array}{r} 181 \\ +\ 765 \\ \hline \end{array}
\qquad
\begin{array}{r} 148 \\ +\ 205 \\ \hline \end{array}
\qquad
\begin{array}{r} 891 \\ +\ 622 \\ \hline \end{array}
$$

D
$$
\begin{array}{r} 732 \\ +\ 699 \\ \hline \end{array}
\qquad
\begin{array}{r} 524 \\ +\ 193 \\ \hline \end{array}
\qquad
\begin{array}{r} 835 \\ +\ 711 \\ \hline \end{array}
\qquad
\begin{array}{r} 570 \\ +\ 572 \\ \hline \end{array}
\qquad
\begin{array}{r} 195 \\ +\ 633 \\ \hline \end{array}
$$

E
$$
\begin{array}{r} 990 \\ +\ 365 \\ \hline \end{array}
\qquad
\begin{array}{r} 445 \\ +\ 629 \\ \hline \end{array}
\qquad
\begin{array}{r} 540 \\ +\ 117 \\ \hline \end{array}
\qquad
\begin{array}{r} 628 \\ +\ 342 \\ \hline \end{array}
\qquad
\begin{array}{r} 503 \\ +\ 272 \\ \hline \end{array}
$$

F
$$
\begin{array}{r} 556 \\ +\ 551 \\ \hline \end{array}
\qquad
\begin{array}{r} 808 \\ +\ 346 \\ \hline \end{array}
\qquad
\begin{array}{r} 920 \\ +\ 247 \\ \hline \end{array}
\qquad
\begin{array}{r} 229 \\ +\ 306 \\ \hline \end{array}
\qquad
\begin{array}{r} 176 \\ +\ 954 \\ \hline \end{array}
$$

**A**

| 803 | 182 | 887 | 177 | 836 |
|-----|-----|-----|-----|-----|
| + 353 | + 907 | + 814 | + 479 | + 292 |

**B**

| 126 | 403 | 163 | 868 | 756 |
|-----|-----|-----|-----|-----|
| + 254 | + 413 | + 846 | + 137 | + 275 |

**C**

| 908 | 942 | 582 | 341 | 793 |
|-----|-----|-----|-----|-----|
| + 550 | + 940 | + 223 | + 372 | + 170 |

**D**

| 165 | 464 | 791 | 655 | 939 |
|-----|-----|-----|-----|-----|
| + 508 | + 140 | + 464 | + 722 | + 807 |

**E**

| 383 | 927 | 250 | 674 | 786 |
|-----|-----|-----|-----|-----|
| + 444 | + 517 | + 487 | + 611 | + 587 |

**F**

| 704 | 148 | 880 | 750 | 913 |
|-----|-----|-----|-----|-----|
| + 309 | + 381 | + 355 | + 127 | + 358 |

A
```
  380        132        622        211        864
+ 726      + 202      + 631      + 361      + 767
```

B
```
  598        747        853        523        350
+ 507      + 980      + 623      + 813      + 605
```

C
```
  917        882        421        742        920
+ 763      + 197      + 525      + 262      + 274
```

D
```
  752        687        416        436        601
+ 959      + 731      + 179      + 839      + 527
```

E
```
  647        460        962        609        814
+ 754      + 729      + 438      + 999      + 244
```

F
```
  998        308        748        643        255
+ 458      + 828      + 740      + 920      + 503
```

A
```
  325       851       691       162       830
+ 735     + 848     + 668     + 375     + 805
```

B
```
  769       485       413       179       949
+ 115     + 118     + 922     + 335     + 651
```

C
```
  969       724       118       193       783
+ 151     + 446     + 226     + 657     + 461
```

D
```
  531       633       462       275       694
+ 666     + 693     + 864     + 881     + 237
```

E
```
  679       704       823       863       555
+ 290     + 801     + 340     + 625     + 134
```

F
```
  337       700       764       292       854
+ 713     + 789     + 300     + 906     + 491
```

A
478        322        767        706        745
+ 774      + 293      + 728      + 994      + 614

B
759        124        728        526        442
+ 899      + 597      + 565      + 521      + 409

C
198        658        856        758        114
+ 810      + 366      + 581      + 720      + 339

D
931        485        727        374        603
+ 893      + 836      + 710      + 721      + 299

E
626        414        248        437        921
+ 105      + 227      + 921      + 119      + 354

F
889        892        676        563        433
+ 600      + 674      + 113      + 347      + 669

A
```
  665      346      904      629      614
+ 575    + 891    + 204    + 680    + 322
```

B
```
  394      715      737      301      321
+ 149    + 827    + 235    + 534    + 384
```

C
```
  344      408      725      701      634
+ 595    + 577    + 131    + 691    + 800
```

D
```
  464      280      302      844      664
+ 885    + 758    + 302    + 775    + 114
```

E
```
  451      275      853      183      146
+ 396    + 442    + 141    + 456    + 416
```

F
```
  978      793      472      547      334
+ 500    + 323    + 523    + 483    + 311
```

A

| 274 | 990 | 960 | 326 | 430 |
|-----|-----|-----|-----|-----|
| + 671 | + 362 | + 984 | + 163 | + 617 |

B

| 177 | 516 | 195 | 292 | 234 |
|-----|-----|-----|-----|-----|
| + 914 | + 451 | + 273 | + 382 | + 819 |

C

| 790 | 597 | 294 | 785 | 111 |
|-----|-----|-----|-----|-----|
| + 843 | + 663 | + 186 | + 287 | + 580 |

D

| 510 | 574 | 435 | 288 | 661 |
|-----|-----|-----|-----|-----|
| + 951 | + 312 | + 203 | + 158 | + 139 |

E

| 905 | 247 | 418 | 122 | 941 |
|-----|-----|-----|-----|-----|
| + 495 | + 676 | + 697 | + 418 | + 389 |

F

| 876 | 485 | 635 | 972 | 654 |
|-----|-----|-----|-----|-----|
| + 977 | + 782 | + 943 | + 154 | + 101 |

A
```
  340      498      499      492      290
+ 880    + 270    + 832    + 753    + 593
```

B
```
  985      507      195      439      411
+ 908    + 583    + 997    + 571    + 178
```

C
```
  976      192      778      848      701
+ 755    + 809    + 705    + 174    + 866
```

D
```
  307      998      448      235      759
+ 719    + 872    + 224    + 752    + 641
```

E
```
  136      642      998      194      641
+ 400    + 443    + 923    + 267    + 978
```

F
```
  352      824      148      402      660
+ 122    + 654    + 338    + 108    + 135
```

A
| 696 | 463 | 734 | 918 | 206 |
| + 776 | + 616 | + 988 | + 639 | + 986 |

B
| 243 | 723 | 264 | 318 | 105 |
| + 207 | + 480 | + 956 | + 910 | + 214 |

C
| 782 | 717 | 447 | 895 | 797 |
| + 165 | + 877 | + 909 | + 148 | + 406 |

D
| 389 | 517 | 125 | 578 | 262 |
| + 737 | + 967 | + 926 | + 928 | + 433 |

E
| 860 | 767 | 744 | 396 | 746 |
| + 276 | + 985 | + 314 | + 518 | + 621 |

F
| 771 | 133 | 253 | 352 | 272 |
| + 337 | + 700 | + 738 | + 714 | + 992 |

A
```
   343        582        130        359        891
 + 350      + 863      + 642      + 794      + 359
```

B
```
   948        107        120        586        958
 + 977      + 461      + 744      + 507      + 273
```

C
```
   846        781        883        262        564
 + 658      + 215      + 978      + 302      + 641
```

D
```
   569        369        415        396        427
 + 467      + 618      + 314      + 532      + 971
```

E
```
   253        173        826        387        441
 + 502      + 567      + 634      + 253      + 596
```

F
```
   242        497        686        564        382
 + 218      + 931      + 606      + 136      + 370
```

**A**

| 638 | 415 | 140 | 680 | 805 |
|---|---|---|---|---|
| + 538 | + 705 | + 690 | + 960 | + 418 |

**B**

| 496 | 247 | 523 | 676 | 436 |
|---|---|---|---|---|
| + 465 | + 873 | + 299 | + 390 | + 802 |

**C**

| 799 | 309 | 198 | 849 | 346 |
|---|---|---|---|---|
| + 628 | + 692 | + 312 | + 930 | + 620 |

**D**

| 990 | 208 | 786 | 491 | 493 |
|---|---|---|---|---|
| + 668 | + 354 | + 506 | + 882 | + 642 |

**E**

| 483 | 419 | 606 | 111 | 794 |
|---|---|---|---|---|
| + 932 | + 158 | + 610 | + 826 | + 154 |

**F**

| 239 | 575 | 583 | 831 | 580 |
|---|---|---|---|---|
| + 786 | + 368 | + 313 | + 114 | + 688 |

**A**

```
   344        298        484        829        559
 + 532      + 363      + 365      + 219      + 449
 ─────      ─────      ─────      ─────      ─────
```

**B**

```
   559        679        941        217        339
 + 962      + 852      + 998      + 515      + 561
 ─────      ─────      ─────      ─────      ─────
```

**C**

```
   164        460        254        444        770
 + 652      + 308      + 288      + 143      + 870
 ─────      ─────      ─────      ─────      ─────
```

**D**

```
   152        381        702        193        506
 + 521      + 182      + 494      + 125      + 888
 ─────      ─────      ─────      ─────      ─────
```

**E**

```
   341        416        411        227        562
 + 566      + 895      + 257      + 856      + 907
 ─────      ─────      ─────      ─────      ─────
```

**F**

```
   375        529        381        445        390
 + 821      + 250      + 318      + 458      + 896
 ─────      ─────      ─────      ─────      ─────
```

**A**

| 897 | 769 | 123 | 105 | 801 |
|-----|-----|-----|-----|-----|
| + 867 | + 416 | + 837 | + 825 | + 708 |

**B**

| 124 | 868 | 203 | 310 | 682 |
|-----|-----|-----|-----|-----|
| + 302 | + 724 | + 345 | + 792 | + 118 |

**C**

| 618 | 248 | 967 | 386 | 295 |
|-----|-----|-----|-----|-----|
| + 231 | + 436 | + 794 | + 297 | + 630 |

**D**

| 976 | 206 | 705 | 915 | 217 |
|-----|-----|-----|-----|-----|
| + 920 | + 800 | + 569 | + 536 | + 205 |

**E**

| 787 | 262 | 861 | 895 | 723 |
|-----|-----|-----|-----|-----|
| + 117 | + 370 | + 220 | + 634 | + 273 |

**F**

| 660 | 997 | 477 | 232 | 929 |
|-----|-----|-----|-----|-----|
| + 693 | + 391 | + 281 | + 404 | + 329 |

**A**

$$\begin{array}{r} 482 \\ + 255 \\ \hline \end{array}$$
$$\begin{array}{r} 472 \\ + 994 \\ \hline \end{array}$$
$$\begin{array}{r} 660 \\ + 928 \\ \hline \end{array}$$
$$\begin{array}{r} 682 \\ + 206 \\ \hline \end{array}$$
$$\begin{array}{r} 673 \\ + 980 \\ \hline \end{array}$$

**B**

$$\begin{array}{r} 770 \\ + 189 \\ \hline \end{array}$$
$$\begin{array}{r} 562 \\ + 489 \\ \hline \end{array}$$
$$\begin{array}{r} 342 \\ + 705 \\ \hline \end{array}$$
$$\begin{array}{r} 358 \\ + 413 \\ \hline \end{array}$$
$$\begin{array}{r} 624 \\ + 294 \\ \hline \end{array}$$

**C**

$$\begin{array}{r} 262 \\ + 524 \\ \hline \end{array}$$
$$\begin{array}{r} 495 \\ + 567 \\ \hline \end{array}$$
$$\begin{array}{r} 529 \\ + 331 \\ \hline \end{array}$$
$$\begin{array}{r} 719 \\ + 530 \\ \hline \end{array}$$
$$\begin{array}{r} 573 \\ + 523 \\ \hline \end{array}$$

**D**

$$\begin{array}{r} 489 \\ + 902 \\ \hline \end{array}$$
$$\begin{array}{r} 625 \\ + 950 \\ \hline \end{array}$$
$$\begin{array}{r} 244 \\ + 271 \\ \hline \end{array}$$
$$\begin{array}{r} 191 \\ + 265 \\ \hline \end{array}$$
$$\begin{array}{r} 871 \\ + 147 \\ \hline \end{array}$$

**E**

$$\begin{array}{r} 705 \\ + 510 \\ \hline \end{array}$$
$$\begin{array}{r} 948 \\ + 807 \\ \hline \end{array}$$
$$\begin{array}{r} 473 \\ + 100 \\ \hline \end{array}$$
$$\begin{array}{r} 307 \\ + 679 \\ \hline \end{array}$$
$$\begin{array}{r} 446 \\ + 736 \\ \hline \end{array}$$

**F**

$$\begin{array}{r} 686 \\ + 371 \\ \hline \end{array}$$
$$\begin{array}{r} 458 \\ + 508 \\ \hline \end{array}$$
$$\begin{array}{r} 323 \\ + 448 \\ \hline \end{array}$$
$$\begin{array}{r} 386 \\ + 721 \\ \hline \end{array}$$
$$\begin{array}{r} 881 \\ + 140 \\ \hline \end{array}$$

A
```
  138        800        424        424        151
+ 585      + 110      + 781      + 641      + 756
```

B
```
  238        395        455        549        592
+ 483      + 924      + 731      + 386      + 274
```

C
```
  770        425        673        812        108
+ 129      + 559      + 915      + 653      + 434
```

D
```
  552        803        559        584        287
+ 168      + 667      + 209      + 103      + 142
```

E
```
  413        676        418        710        224
+ 845      + 811      + 969      + 392      + 462
```

F
```
  129        210        431        821        193
+ 303      + 555      + 846      + 422      + 287
```

A

$$957 + 685$$    $$510 + 843$$    $$994 + 565$$    $$363 + 452$$    $$336 + 803$$

B

$$610 + 587$$    $$462 + 243$$    $$527 + 842$$    $$680 + 578$$    $$463 + 573$$

C

$$337 + 460$$    $$374 + 492$$    $$531 + 488$$    $$730 + 680$$    $$863 + 952$$

D

$$311 + 267$$    $$143 + 689$$    $$173 + 956$$    $$120 + 517$$    $$439 + 806$$

E

$$231 + 838$$    $$536 + 479$$    $$837 + 951$$    $$122 + 455$$    $$133 + 453$$

F

$$914 + 137$$    $$721 + 410$$    $$607 + 631$$    $$463 + 702$$    $$787 + 376$$

A
|  | 688 | 636 | 219 | 502 | 611 |
|--|-----|-----|-----|-----|-----|
|  | + 184 | + 592 | + 909 | + 401 | + 238 |

B
|  | 777 | 540 | 381 | 637 | 164 |
|--|-----|-----|-----|-----|-----|
|  | + 196 | + 593 | + 855 | + 682 | + 942 |

C
|  | 547 | 136 | 464 | 648 | 654 |
|--|-----|-----|-----|-----|-----|
|  | + 145 | + 706 | + 396 | + 836 | + 473 |

D
|  | 429 | 779 | 227 | 261 | 660 |
|--|-----|-----|-----|-----|-----|
|  | + 752 | + 263 | + 990 | + 172 | + 169 |

E
|  | 449 | 346 | 963 | 415 | 338 |
|--|-----|-----|-----|-----|-----|
|  | + 865 | + 829 | + 152 | + 655 | + 296 |

F
|  | 117 | 269 | 280 | 272 | 636 |
|--|-----|-----|-----|-----|-----|
|  | + 138 | + 307 | + 472 | + 381 | + 577 |

Day:  19

Name:

Date:

Time:          :

Score:        /30

Rating:  ☆☆☆☆☆

A
```
  573        688        576        337        288
+ 269      + 333      + 614      + 783      + 768
```

B
```
  853        510        803        340        906
+ 407      + 520      + 629      + 435      + 388
```

C
```
  513        840        462        723        954
+ 726      + 604      + 780      + 801      + 623
```

D
```
  158        460        721        465        722
+ 626      + 545      + 779      + 827      + 851
```

E
```
  823        947        498        978        485
+ 120      + 159      + 447      + 463      + 348
```

F
```
  875        606        991        531        159
+ 208      + 863      + 973      + 983      + 877
```

A
```
   599      586      179      133      632
 + 709    + 400    + 910    + 787    + 989
```

B
```
   921      527      832      962      943
 + 996    + 306    + 384    + 883    + 372
```

C
```
   233      200      898      422      992
 + 695    + 153    + 872    + 584    + 264
```

D
```
   337      592      254      922      259
 + 123    + 824    + 600    + 651    + 499
```

E
```
   327      381      991      509      931
 + 311    + 501    + 606    + 660    + 240
```

F
```
   315      160      462      631      991
 + 346    + 828    + 415    + 207    + 995
```

A
```
  511        153        623        768        732
+ 474      + 525      + 343      + 899      + 300
```

B
```
  634        372        715        666        548
+ 179      + 908      + 534      + 849      + 461
```

C
```
  266        281        793        485        515
+ 262      + 176      + 421      + 616      + 696
```

D
```
  683        391        878        324        319
+ 815      + 210      + 468      + 757      + 526
```

E
```
  860        272        668        878        132
+ 115      + 914      + 957      + 495      + 445
```

F
```
  232        333        877        399        444
+ 588      + 912      + 389      + 844      + 355
```

A
```
  377        971        399        987        526
+ 647      + 961      + 681      + 663      + 596
```

B
```
  969        439        240        739        178
+ 200      + 818      + 442      + 744      + 658
```

C
```
  806        304        106        995        385
+ 929      + 411      + 546      + 617      + 551
```

D
```
  767        892        804        761        352
+ 598      + 101      + 949      + 793      + 190
```

E
```
  191        532        723        798        379
+ 202      + 612      + 886      + 379      + 380
```

F
```
  811        575        204        385        606
+ 500      + 105      + 423      + 599      + 820
```

A

| 589 | 899 | 700 | 349 | 978 |
|-----|-----|-----|-----|-----|
| + 562 | + 275 | + 734 | + 749 | + 289 |

B

| 585 | 865 | 743 | 913 | 524 |
|-----|-----|-----|-----|-----|
| + 514 | + 903 | + 777 | + 323 | + 237 |

C

| 458 | 569 | 928 | 138 | 166 |
|-----|-----|-----|-----|-----|
| + 893 | + 450 | + 174 | + 285 | + 186 |

D

| 476 | 630 | 374 | 518 | 983 |
|-----|-----|-----|-----|-----|
| + 982 | + 134 | + 889 | + 775 | + 550 |

E

| 970 | 900 | 541 | 924 | 123 |
|-----|-----|-----|-----|-----|
| + 782 | + 417 | + 228 | + 486 | + 840 |

F

| 553 | 657 | 817 | 107 | 900 |
|-----|-----|-----|-----|-----|
| + 232 | + 310 | + 295 | + 225 | + 324 |

A

| 483 | 901 | 386 | 815 | 496 |
|---|---|---|---|---|
| + 798 | + 369 | + 970 | + 831 | + 813 |

B

| 558 | 242 | 151 | 654 | 295 |
|---|---|---|---|---|
| + 471 | + 213 | + 645 | + 244 | + 861 |

C

| 493 | 237 | 632 | 656 | 464 |
|---|---|---|---|---|
| + 247 | + 712 | + 897 | + 745 | + 239 |

D

| 883 | 335 | 836 | 141 | 979 |
|---|---|---|---|---|
| + 991 | + 677 | + 314 | + 135 | + 466 |

E

| 446 | 760 | 198 | 999 | 937 |
|---|---|---|---|---|
| + 941 | + 839 | + 357 | + 866 | + 850 |

F

| 969 | 775 | 705 | 832 | 642 |
|---|---|---|---|---|
| + 894 | + 132 | + 187 | + 977 | + 519 |

**A**

| | | | | |
| --- | --- | --- | --- | --- |
| 485<br>+ 227 | 387<br>+ 282 | 221<br>+ 378 | 539<br>+ 198 | 157<br>+ 675 |

**B**

| | | | | |
| --- | --- | --- | --- | --- |
| 435<br>+ 940 | 534<br>+ 344 | 894<br>+ 173 | 194<br>+ 797 | 473<br>+ 424 |

**C**

| | | | | |
| --- | --- | --- | --- | --- |
| 270<br>+ 618 | 556<br>+ 812 | 628<br>+ 804 | 438<br>+ 770 | 378<br>+ 694 |

**D**

| | | | | |
| --- | --- | --- | --- | --- |
| 371<br>+ 480 | 531<br>+ 317 | 373<br>+ 224 | 778<br>+ 509 | 171<br>+ 332 |

**E**

| | | | | |
| --- | --- | --- | --- | --- |
| 900<br>+ 529 | 945<br>+ 808 | 741<br>+ 635 | 226<br>+ 597 | 633<br>+ 983 |

**F**

| | | | | |
| --- | --- | --- | --- | --- |
| 773<br>+ 228 | 906<br>+ 601 | 372<br>+ 722 | 280<br>+ 970 | 728<br>+ 230 |

A

```
  122        672        689        717        716
+ 399      + 371      + 131      + 596      + 603
```

B

```
  228        671        348        976        793
+ 635      + 337      + 457      + 458      + 110
```

C

```
  297        778        742        411        102
+ 393      + 965      + 927      + 869      + 451
```

D

```
  670        721        672        917        134
+ 760      + 164      + 719      + 825      + 240
```

E

```
  321        724        408        160        569
+ 899      + 726      + 430      + 904      + 187
```

F

```
  777        641        413        272        287
+ 754      + 789      + 372      + 464      + 660
```

**A**

```
   953        760        215        436        875
 + 253      + 876      + 797      + 755      + 268
 _____     _____     _____     _____     _____
```

**B**

```
   891        256        961        313        273
 + 514      + 284      + 188      + 263      + 671
 _____     _____     _____     _____     _____
```

**C**

```
   143        695        450        366        535
 + 565      + 685      + 321      + 170      + 710
 _____     _____     _____     _____     _____
```

**D**

```
   599        861        506        434        349
 + 493      + 261      + 239      + 317      + 959
 _____     _____     _____     _____     _____
```

**E**

```
   201        206        299        279        438
 + 103      + 324      + 796      + 993      + 872
 _____     _____     _____     _____     _____
```

**F**

```
   698        897        153        624        583
 + 304      + 559      + 728      + 305      + 155
 _____     _____     _____     _____     _____
```

A
267
+ 884

213
+ 222

468
+ 790

354
+ 166

463
+ 889

B
456
+ 557

788
+ 613

413
+ 693

458
+ 945

314
+ 553

C
979
+ 606

101
+ 499

706
+ 316

739
+ 717

579
+ 202

D
912
+ 788

479
+ 962

345
+ 779

542
+ 771

553
+ 579

E
154
+ 979

458
+ 116

335
+ 581

182
+ 854

614
+ 126

F
391
+ 757

754
+ 403

389
+ 364

790
+ 550

930
+ 416

A
$$577 + 214$$  $$654 + 154$$  $$640 + 617$$  $$528 + 515$$  $$503 + 231$$

B
$$143 + 489$$  $$922 + 497$$  $$551 + 644$$  $$523 + 282$$  $$740 + 570$$

C
$$456 + 458$$  $$510 + 615$$  $$757 + 264$$  $$407 + 397$$  $$384 + 427$$

D
$$845 + 377$$  $$526 + 485$$  $$980 + 142$$  $$407 + 402$$  $$741 + 994$$

E
$$209 + 847$$  $$628 + 217$$  $$554 + 677$$  $$787 + 819$$  $$191 + 941$$

F
$$852 + 902$$  $$216 + 833$$  $$677 + 176$$  $$821 + 393$$  $$298 + 134$$

A
```
  546       908       242       969       778
+ 939     + 442     + 639     + 287     + 307
```

B
```
  878       167       541       668       543
+ 151     + 494     + 817     + 758     + 258
```

C
```
  145       463       603       261       536
+ 126     + 912     + 481     + 590     + 526
```

D
```
  557       843       315       597       671
+ 740     + 582     + 852     + 143     + 924
```

E
```
  393       381       595       675       214
+ 709     + 171     + 127     + 311     + 527
```

F
```
  816       588       341       582       559
+ 780     + 505     + 766     + 870     + 220
```

A
$$311 + 296$$
$$164 + 553$$
$$493 + 730$$
$$490 + 744$$
$$353 + 293$$

B
$$404 + 191$$
$$456 + 815$$
$$185 + 904$$
$$548 + 880$$
$$181 + 580$$

C
$$969 + 658$$
$$763 + 781$$
$$601 + 927$$
$$673 + 343$$
$$558 + 466$$

D
$$452 + 299$$
$$901 + 417$$
$$490 + 978$$
$$659 + 915$$
$$326 + 761$$

E
$$602 + 595$$
$$991 + 792$$
$$280 + 504$$
$$368 + 890$$
$$804 + 998$$

F
$$964 + 549$$
$$630 + 641$$
$$931 + 588$$
$$206 + 630$$
$$947 + 857$$

A
```
  644        407        725        957        757
+ 541      + 746      + 408      + 395      + 801
```

B
```
  589        828        240        820        233
+ 565      + 401      + 823      + 550      + 381
```

C
```
  473        711        632        675        698
+ 185      + 365      + 811      + 748      + 919
```

D
```
  475        536        665        328        981
+ 430      + 731      + 276      + 294      + 574
```

E
```
  292        325        819        943        490
+ 885      + 850      + 316      + 888      + 271
```

F
```
  243        442        422        322        474
+ 831      + 233      + 965      + 903      + 589
```

30

60

A
| 350 | 298 | 251 | 902 | 992 |
| + 832 | + 383 | + 959 | + 159 | + 357 |

B
| 325 | 927 | 712 | 611 | 170 |
| + 217 | + 126 | + 254 | + 241 | + 115 |

C
| 145 | 596 | 921 | 734 | 373 |
| + 346 | + 872 | + 997 | + 107 | + 807 |

D
| 694 | 481 | 206 | 156 | 320 |
| + 150 | + 434 | + 719 | + 751 | + 442 |

E
| 562 | 552 | 391 | 217 | 981 |
| + 287 | + 772 | + 628 | + 461 | + 505 |

F
| 117 | 489 | 262 | 373 | 704 |
| + 330 | + 864 | + 548 | + 389 | + 167 |

A
```
  509        550        726        792        402
+ 473      + 973      + 626      + 755      + 247
```

B
```
  414        256        147        721        959
+ 639      + 779      + 572      + 817      + 843
```

C
```
  447        158        147        187        266
+ 180      + 735      + 310      + 305      + 528
```

D
```
  107        626        401        336        862
+ 144      + 561      + 983      + 789      + 309
```

E
```
  518        832        825        977        848
+ 447      + 470      + 891      + 958      + 476
```

F
```
  132        239        722        123        123
+ 782      + 671      + 594      + 770      + 410
```

A

$$717 + 821$$   $$882 + 477$$   $$334 + 132$$   $$972 + 559$$   $$217 + 942$$

B

$$299 + 213$$   $$995 + 155$$   $$833 + 950$$   $$188 + 889$$   $$940 + 962$$

C

$$815 + 479$$   $$431 + 375$$   $$733 + 780$$   $$621 + 489$$   $$627 + 941$$

D

$$847 + 814$$   $$832 + 638$$   $$625 + 703$$   $$839 + 940$$   $$653 + 764$$

E

$$573 + 187$$   $$777 + 257$$   $$945 + 396$$   $$647 + 861$$   $$417 + 546$$

F

$$696 + 741$$   $$985 + 988$$   $$636 + 977$$   $$838 + 359$$   $$782 + 579$$

A

| 319 | 446 | 722 | 352 | 374 |
|-----|-----|-----|-----|-----|
| + 724 | + 264 | + 863 | + 260 | + 204 |

B

| 844 | 263 | 633 | 846 | 721 |
|-----|-----|-----|-----|-----|
| + 266 | + 361 | + 390 | + 987 | + 605 |

C

| 450 | 874 | 811 | 137 | 924 |
|-----|-----|-----|-----|-----|
| + 252 | + 722 | + 174 | + 787 | + 778 |

D

| 224 | 123 | 755 | 322 | 130 |
|-----|-----|-----|-----|-----|
| + 547 | + 900 | + 679 | + 539 | + 895 |

E

| 708 | 114 | 526 | 667 | 746 |
|-----|-----|-----|-----|-----|
| + 846 | + 421 | + 696 | + 369 | + 926 |

F

| 503 | 645 | 380 | 758 | 532 |
|-----|-----|-----|-----|-----|
| + 169 | + 273 | + 975 | + 830 | + 842 |

**A**

| 834 | 313 | 442 | 828 | 727 |
|-----|-----|-----|-----|-----|
| + 131 | + 388 | + 822 | + 769 | + 600 |

**B**

| 813 | 998 | 373 | 764 | 970 |
|-----|-----|-----|-----|-----|
| + 802 | + 631 | + 957 | + 177 | + 425 |

**C**

| 457 | 561 | 182 | 618 | 131 |
|-----|-----|-----|-----|-----|
| + 128 | + 627 | + 491 | + 469 | + 200 |

**D**

| 391 | 488 | 990 | 330 | 469 |
|-----|-----|-----|-----|-----|
| + 582 | + 835 | + 458 | + 216 | + 532 |

**E**

| 377 | 267 | 635 | 862 | 588 |
|-----|-----|-----|-----|-----|
| + 610 | + 664 | + 199 | + 845 | + 298 |

**F**

| 903 | 676 | 783 | 939 | 639 |
|-----|-----|-----|-----|-----|
| + 865 | + 289 | + 897 | + 623 | + 920 |

A
```
  495        877        682        850        875
+ 667      + 342      + 481      + 314      + 524
```

B
```
  756        530        384        917        284
+ 984      + 143      + 697      + 938      + 604
```

C
```
  146        396        512        300        106
+ 327      + 513      + 790      + 147      + 683
```

D
```
  638        473        394        525        194
+ 402      + 352      + 320      + 445      + 971
```

E
```
  232        853        843        774        763
+ 720      + 229      + 483      + 157      + 475
```

F
```
  397        558        861        970        623
+ 462      + 791      + 211      + 256      + 500
```

| | Day: | 39 | | Date: | | | Score: | /30 |
| --- | --- | --- | --- | --- | --- | --- | --- | --- |
| | Name: | | | Time: | : | | Rating: | ☆☆☆☆☆ |

A

```
  543        734        482        685        477
+ 313      + 106      + 974      + 961      + 745
-----      -----      -----      -----      -----
```

B

```
  788        146        450        928        181
+ 426      + 270      + 603      + 378      + 837
-----      -----      -----      -----      -----
```

C

```
  367        499        472        507        209
+ 952      + 632      + 231      + 555      + 912
-----      -----      -----      -----      -----
```

D

```
  517        904        435        555        619
+ 478      + 972      + 487      + 621      + 182
-----      -----      -----      -----      -----
```

E

```
  640        847        178        922        490
+ 160      + 317      + 808      + 615      + 319
-----      -----      -----      -----      -----
```

F

```
  447        555        386        761        434
+ 129      + 866      + 183      + 441      + 987
-----      -----      -----      -----      -----
```

Day: 40

Name:

Date:

Time:      :

Score:      /30

Rating: ☆☆☆☆☆

A
| 278 | 190 | 465 | 894 | 930 |
|-----|-----|-----|-----|-----|
| + 149 | + 907 | + 873 | + 297 | + 401 |

B
| 338 | 443 | 172 | 240 | 525 |
|-----|-----|-----|-----|-----|
| + 972 | + 271 | + 724 | + 842 | + 656 |

C
| 289 | 974 | 595 | 304 | 291 |
|-----|-----|-----|-----|-----|
| + 312 | + 429 | + 176 | + 408 | + 204 |

D
| 925 | 336 | 618 | 501 | 614 |
|-----|-----|-----|-----|-----|
| + 585 | + 306 | + 875 | + 562 | + 699 |

E
| 550 | 235 | 523 | 865 | 206 |
|-----|-----|-----|-----|-----|
| + 435 | + 524 | + 806 | + 248 | + 132 |

F
| 941 | 235 | 671 | 926 | 937 |
|-----|-----|-----|-----|-----|
| + 789 | + 265 | + 800 | + 977 | + 961 |

A
```
  392        281        768        748        612
+ 763      + 668      + 660      + 879      + 227
```

B
```
  292        105        352        846        506
+ 563      + 374      + 851      + 532      + 346
```

C
```
  133        157        420        587        235
+ 419      + 598      + 745      + 138      + 911
```

D
```
  524        725        325        891        915
+ 573      + 516      + 734      + 234      + 391
```

E
```
  278        808        273        218        848
+ 867      + 124      + 721      + 329      + 447
```

F
```
  134        381        230        804        251
+ 621      + 578      + 512      + 155      + 966
```

**A**

| 365 | 101 | 814 | 727 | 463 |
|-----|-----|-----|-----|-----|
| + 474 | + 528 | + 828 | + 191 | + 854 |

**B**

| 424 | 880 | 958 | 204 | 821 |
|-----|-----|-----|-----|-----|
| + 853 | + 531 | + 684 | + 328 | + 384 |

**C**

| 893 | 645 | 811 | 726 | 545 |
|-----|-----|-----|-----|-----|
| + 910 | + 122 | + 831 | + 743 | + 748 |

**D**

| 935 | 406 | 278 | 763 | 993 |
|-----|-----|-----|-----|-----|
| + 877 | + 954 | + 529 | + 926 | + 453 |

**E**

| 725 | 440 | 820 | 502 | 376 |
|-----|-----|-----|-----|-----|
| + 751 | + 354 | + 333 | + 535 | + 691 |

**F**

| 205 | 286 | 822 | 117 | 638 |
|-----|-----|-----|-----|-----|
| + 868 | + 804 | + 576 | + 909 | + 848 |

Day: 43

Date:

Score: /30

Name:

Time: :

Rating: ☆☆☆☆☆

**A**

| 447 | 709 | 664 | 715 | 871 |
|-----|-----|-----|-----|-----|
| + 327 | + 352 | + 759 | + 822 | + 518 |

**B**

| 358 | 750 | 533 | 533 | 602 |
|-----|-----|-----|-----|-----|
| + 223 | + 767 | + 195 | + 670 | + 645 |

**C**

| 951 | 949 | 577 | 373 | 395 |
|-----|-----|-----|-----|-----|
| + 637 | + 658 | + 950 | + 837 | + 653 |

**D**

| 660 | 524 | 768 | 190 | 445 |
|-----|-----|-----|-----|-----|
| + 421 | + 701 | + 287 | + 148 | + 888 |

**E**

| 678 | 979 | 535 | 469 | 325 |
|-----|-----|-----|-----|-----|
| + 788 | + 185 | + 568 | + 795 | + 916 |

**F**

| 932 | 329 | 818 | 591 | 281 |
|-----|-----|-----|-----|-----|
| + 120 | + 282 | + 440 | + 770 | + 549 |

A
$$667 + 241$$
$$876 + 235$$
$$559 + 862$$
$$880 + 915$$
$$886 + 192$$

B
$$657 + 581$$
$$638 + 206$$
$$773 + 957$$
$$636 + 651$$
$$854 + 156$$

C
$$409 + 749$$
$$449 + 503$$
$$930 + 111$$
$$527 + 636$$
$$685 + 760$$

D
$$933 + 631$$
$$678 + 914$$
$$139 + 814$$
$$973 + 510$$
$$198 + 951$$

E
$$896 + 648$$
$$386 + 314$$
$$199 + 523$$
$$254 + 819$$
$$554 + 405$$

F
$$274 + 777$$
$$135 + 445$$
$$458 + 712$$
$$932 + 280$$
$$799 + 646$$

A
```
  303        570        805        529        710
+ 980      + 825      + 849      + 996      + 442
```

B
```
  573        171        170        825        562
+ 200      + 723      + 971      + 774      + 478
```

C
```
  786        958        876        580        792
+ 905      + 588      + 839      + 464      + 173
```

D
```
  337        393        629        743        594
+ 221      + 584      + 444      + 418      + 190
```

E
```
  312        372        517        427        536
+ 309      + 714      + 681      + 245      + 892
```

F
```
  261        585        634        305        284
+ 553      + 566      + 509      + 592      + 520
```

A
788 + 308
683 + 619
570 + 948
625 + 872
962 + 973

B
567 + 514
917 + 861
314 + 693
536 + 153
784 + 466

C
947 + 321
155 + 771
907 + 291
504 + 719
526 + 990

D
286 + 605
901 + 385
250 + 537
720 + 259
236 + 292

E
879 + 513
199 + 446
771 + 296
448 + 211
169 + 552

F
416 + 104
447 + 666
702 + 832
201 + 764
294 + 114

A
| 738 | 628 | 300 | 981 | 673 |
|---|---|---|---|---|
| + 705 | + 608 | + 147 | + 439 | + 360 |

B
| 219 | 776 | 626 | 213 | 541 |
|---|---|---|---|---|
| + 611 | + 281 | + 762 | + 533 | + 174 |

C
| 848 | 893 | 874 | 534 | 339 |
|---|---|---|---|---|
| + 593 | + 918 | + 473 | + 479 | + 402 |

D
| 884 | 548 | 221 | 901 | 273 |
|---|---|---|---|---|
| + 663 | + 538 | + 896 | + 129 | + 243 |

E
| 679 | 400 | 939 | 253 | 239 |
|---|---|---|---|---|
| + 160 | + 625 | + 884 | + 986 | + 942 |

F
| 122 | 475 | 182 | 740 | 776 |
|---|---|---|---|---|
| + 267 | + 255 | + 672 | + 655 | + 239 |

Day: 48

Name:

Date:

Time:      :

Score:   /30

Rating: ☆☆☆☆☆

**A**

| 325 | 994 | 959 | 676 | 477 |
|---|---|---|---|---|
| + 895 | + 731 | + 984 | + 276 | + 561 |

**B**

| 188 | 138 | 866 | 807 | 842 |
|---|---|---|---|---|
| + 794 | + 412 | + 534 | + 618 | + 901 |

**C**

| 172 | 373 | 582 | 125 | 624 |
|---|---|---|---|---|
| + 555 | + 417 | + 801 | + 123 | + 674 |

**D**

| 683 | 201 | 432 | 439 | 824 |
|---|---|---|---|---|
| + 998 | + 134 | + 560 | + 164 | + 933 |

**E**

| 283 | 538 | 992 | 361 | 460 |
|---|---|---|---|---|
| + 614 | + 274 | + 569 | + 376 | + 261 |

**F**

| 836 | 380 | 345 | 370 | 158 |
|---|---|---|---|---|
| + 293 | + 781 | + 166 | + 551 | + 782 |

**A**

$$146 + 242$$  $$923 + 443$$  $$876 + 604$$  $$927 + 747$$  $$222 + 113$$

**B**

$$900 + 278$$  $$946 + 830$$  $$869 + 370$$  $$935 + 300$$  $$737 + 591$$

**C**

$$738 + 845$$  $$923 + 386$$  $$227 + 295$$  $$995 + 594$$  $$457 + 679$$

**D**

$$236 + 438$$  $$483 + 400$$  $$514 + 818$$  $$880 + 626$$  $$528 + 923$$

**E**

$$897 + 133$$  $$530 + 677$$  $$981 + 290$$  $$884 + 502$$  $$566 + 469$$

**F**

$$644 + 373$$  $$458 + 889$$  $$620 + 690$$  $$645 + 963$$  $$204 + 128$$

A
| 354 | 383 | 958 | 141 | 346 |
| + 175 | + 142 | + 550 | + 960 | + 755 |

B
| 783 | 769 | 606 | 285 | 847 |
| + 303 | + 649 | + 703 | + 761 | + 565 |

C
| 220 | 764 | 813 | 777 | 966 |
| + 357 | + 728 | + 729 | + 436 | + 407 |

D
| 310 | 475 | 341 | 190 | 900 |
| + 313 | + 451 | + 917 | + 991 | + 495 |

E
| 642 | 756 | 793 | 284 | 153 |
| + 640 | + 791 | + 298 | + 165 | + 628 |

F
| 250 | 349 | 952 | 874 | 977 |
| + 507 | + 711 | + 547 | + 392 | + 434 |

**A**

| 169 | 253 | 292 | 702 | 222 |
|-----|-----|-----|-----|-----|
| + 158 | + 263 | + 660 | + 312 | + 278 |

**B**

| 350 | 957 | 152 | 487 | 754 |
|-----|-----|-----|-----|-----|
| + 508 | + 352 | + 446 | + 841 | + 222 |

**C**

| 706 | 331 | 384 | 756 | 294 |
|-----|-----|-----|-----|-----|
| + 972 | + 632 | + 512 | + 455 | + 967 |

**D**

| 872 | 533 | 122 | 699 | 140 |
|-----|-----|-----|-----|-----|
| + 856 | + 264 | + 872 | + 899 | + 664 |

**E**

| 424 | 536 | 392 | 368 | 502 |
|-----|-----|-----|-----|-----|
| + 902 | + 755 | + 519 | + 500 | + 700 |

**F**

| 495 | 723 | 325 | 749 | 570 |
|-----|-----|-----|-----|-----|
| + 117 | + 366 | + 200 | + 562 | + 720 |

**A**

| 721 | 581 | 851 | 919 | 666 |
|-----|-----|-----|-----|-----|
| − 121 | − 548 | − 441 | − 731 | − 414 |

**B**

| 911 | 828 | 847 | 958 | 562 |
|-----|-----|-----|-----|-----|
| − 678 | − 410 | − 766 | − 696 | − 221 |

**C**

| 783 | 896 | 860 | 750 | 975 |
|-----|-----|-----|-----|-----|
| − 392 | − 680 | − 213 | − 664 | − 750 |

**D**

| 708 | 820 | 729 | 880 | 947 |
|-----|-----|-----|-----|-----|
| − 627 | − 203 | − 291 | − 544 | − 668 |

**E**

| 542 | 801 | 645 | 988 | 643 |
|-----|-----|-----|-----|-----|
| − 157 | − 385 | − 141 | − 476 | − 398 |

**F**

| 706 | 503 | 835 | 899 | 989 |
|-----|-----|-----|-----|-----|
| − 427 | − 472 | − 369 | − 720 | − 637 |

A
```
  565        544        945        669        988
- 500      - 450      - 515      - 540      - 467
```

B
```
  962        794        651        690        948
- 412      - 527      - 490      - 203      - 719
```

C
```
  467        834        288        886        786
- 261      - 340      - 279      - 651      - 167
```

D
```
  849        958        805        967        478
- 217      - 370      - 428      - 882      - 378
```

E
```
  841        992        777        957        944
- 820      - 186      - 146      - 455      - 558
```

F
```
  870        595        859        819        841
- 568      - 240      - 490      - 696      - 267
```

| | | | | |
|---|---|---|---|---|
| A | 768 − 101 | 991 − 782 | 969 − 876 | 955 − 535 | 575 − 262 |

| | | | | |
|---|---|---|---|---|
| B | 539 − 150 | 724 − 246 | 877 − 734 | 896 − 597 | 860 − 766 |

| | | | | |
|---|---|---|---|---|
| C | 461 − 213 | 994 − 268 | 662 − 547 | 787 − 231 | 805 − 429 |

| | | | | |
|---|---|---|---|---|
| D | 921 − 179 | 363 − 207 | 492 − 286 | 722 − 663 | 381 − 369 |

| | | | | |
|---|---|---|---|---|
| E | 567 − 292 | 739 − 715 | 879 − 781 | 767 − 540 | 976 − 891 |

| | | | | |
|---|---|---|---|---|
| F | 457 − 244 | 631 − 460 | 437 − 391 | 890 − 191 | 262 − 130 |

| Day: | 4 | | Date: | | Score: | /30 |
|------|---|--|-------|--|--------|-----|
| Name: | | | Time: | : | Rating: | ☆☆☆☆☆ |

**A**

672
− 465

664
− 416

798
− 431

769
− 295

814
− 775

**B**

689
− 124

867
− 593

927
− 644

661
− 157

530
− 314

**C**

923
− 859

945
− 597

676
− 342

671
− 492

982
− 975

**D**

244
− 112

319
− 225

180
− 166

884
− 335

538
− 496

**E**

868
− 648

722
− 279

985
− 292

296
− 216

588
− 585

**F**

775
− 298

796
− 159

834
− 626

631
− 360

907
− 107

A
$$543 - 434$$
$$917 - 469$$
$$364 - 209$$
$$711 - 196$$
$$287 - 267$$

B
$$967 - 243$$
$$864 - 741$$
$$576 - 303$$
$$855 - 690$$
$$392 - 256$$

C
$$604 - 294$$
$$529 - 359$$
$$842 - 827$$
$$611 - 146$$
$$769 - 732$$

D
$$377 - 287$$
$$745 - 252$$
$$657 - 349$$
$$438 - 337$$
$$840 - 744$$

E
$$716 - 172$$
$$481 - 180$$
$$683 - 302$$
$$295 - 135$$
$$177 - 138$$

F
$$632 - 248$$
$$683 - 279$$
$$838 - 344$$
$$778 - 194$$
$$246 - 163$$

A
| 617 | 898 | 860 | 989 | 763 |
|-----|-----|-----|-----|-----|
| − 462 | − 529 | − 216 | − 703 | − 675 |

B
| 817 | 754 | 895 | 902 | 943 |
|-----|-----|-----|-----|-----|
| − 706 | − 256 | − 841 | − 680 | − 337 |

C
| 866 | 806 | 763 | 753 | 979 |
|-----|-----|-----|-----|-----|
| − 791 | − 659 | − 325 | − 197 | − 426 |

D
| 938 | 717 | 694 | 916 | 973 |
|-----|-----|-----|-----|-----|
| − 790 | − 309 | − 491 | − 568 | − 360 |

E
| 758 | 532 | 853 | 510 | 563 |
|-----|-----|-----|-----|-----|
| − 594 | − 445 | − 527 | − 323 | − 126 |

F
| 435 | 750 | 957 | 754 | 396 |
|-----|-----|-----|-----|-----|
| − 127 | − 685 | − 148 | − 343 | − 251 |

A
| 693 | 647 | 479 | 991 | 796 |
| − 257 | − 590 | − 170 | − 474 | − 301 |

B
| 299 | 655 | 971 | 926 | 966 |
| − 272 | − 611 | − 509 | − 334 | − 286 |

C
| 270 | 929 | 651 | 823 | 707 |
| − 217 | − 652 | − 267 | − 811 | − 159 |

D
| 868 | 976 | 880 | 587 | 931 |
| − 237 | − 692 | − 653 | − 497 | − 771 |

E
| 704 | 798 | 627 | 847 | 410 |
| − 387 | − 155 | − 607 | − 437 | − 179 |

F
| 867 | 912 | 452 | 599 | 962 |
| − 638 | − 423 | − 166 | − 243 | − 766 |

A
```
  888        388        479        893        826
- 845      - 270      - 317      - 778      - 403
```

B
```
  516        888        724        689        759
- 355      - 655      - 431      - 121      - 706
```

C
```
  927        957        830        968        812
- 815      - 429      - 477      - 561      - 330
```

D
```
  948        274        660        728        877
- 755      - 243      - 581      - 442      - 595
```

E
```
  515        384        570        823        973
- 469      - 244      - 185      - 540      - 722
```

F
```
  979        595        486        325        795
- 585      - 322      - 292      - 236      - 327
```

A
| 985 | 769 | 954 | 353 | 539 |
| − 493 | − 634 | − 218 | − 242 | − 380 |

B
| 639 | 861 | 939 | 323 | 265 |
| − 506 | − 156 | − 149 | − 177 | − 192 |

C
| 965 | 553 | 990 | 508 | 997 |
| − 902 | − 486 | − 458 | − 160 | − 684 |

D
| 980 | 324 | 842 | 844 | 219 |
| − 510 | − 124 | − 215 | − 264 | − 134 |

E
| 923 | 502 | 262 | 348 | 574 |
| − 192 | − 107 | − 113 | − 216 | − 102 |

F
| 486 | 360 | 938 | 728 | 340 |
| − 353 | − 193 | − 871 | − 133 | − 156 |

Day: 10

Date:

Score: /30

Name:

Time: :

Rating: ☆☆☆☆☆

A

552
− 429

723
− 548

757
− 363

570
− 431

399
− 208

B

542
− 432

822
− 715

531
− 117

666
− 126

842
− 830

C

566
− 251

700
− 215

718
− 190

497
− 232

457
− 296

D

707
− 351

322
− 164

354
− 163

974
− 909

748
− 412

E

638
− 525

699
− 452

784
− 555

696
− 529

344
− 283

F

822
− 494

864
− 492

536
− 228

975
− 761

569
− 397

30

60

A

```
  840        491        637        622        737
- 746      - 288      - 262      - 110      - 565
```

B

```
  813        691        594        390        701
- 120      - 206      - 391      - 326      - 395
```

C

```
  692        349        733        561        466
- 469      - 183      - 253      - 202      - 335
```

D

```
  415        570        749        407        188
- 126      - 562      - 187      - 225      - 123
```

E

```
  886        627        500        648        730
- 581      - 356      - 451      - 374      - 404
```

F

```
  850        865        407        465        673
- 289      - 456      - 152      - 306      - 199
```

A
| 493 | 960 | 990 | 712 | 713 |
|-----|-----|-----|-----|-----|
| − 334 | − 196 | − 290 | − 271 | − 612 |

B
| 443 | 687 | 881 | 964 | 454 |
|-----|-----|-----|-----|-----|
| − 111 | − 650 | − 789 | − 547 | − 259 |

C
| 939 | 633 | 700 | 964 | 801 |
|-----|-----|-----|-----|-----|
| − 433 | − 401 | − 481 | − 807 | − 645 |

D
| 861 | 449 | 780 | 890 | 910 |
|-----|-----|-----|-----|-----|
| − 632 | − 308 | − 641 | − 225 | − 604 |

E
| 755 | 895 | 929 | 917 | 669 |
|-----|-----|-----|-----|-----|
| − 588 | − 561 | − 644 | − 550 | − 549 |

F
| 829 | 999 | 989 | 524 | 820 |
|-----|-----|-----|-----|-----|
| − 780 | − 830 | − 464 | − 210 | − 545 |

A
| 456 | 856 | 231 | 685 | 958 |
|-----|-----|-----|-----|-----|
| − 179 | − 557 | − 211 | − 549 | − 694 |

B
| 973 | 858 | 910 | 962 | 639 |
|-----|-----|-----|-----|-----|
| − 467 | − 583 | − 623 | − 460 | − 609 |

C
| 913 | 752 | 875 | 328 | 477 |
|-----|-----|-----|-----|-----|
| − 340 | − 670 | − 786 | − 144 | − 247 |

D
| 666 | 756 | 896 | 375 | 974 |
|-----|-----|-----|-----|-----|
| − 536 | − 707 | − 848 | − 251 | − 824 |

E
| 667 | 915 | 997 | 551 | 905 |
|-----|-----|-----|-----|-----|
| − 256 | − 595 | − 994 | − 550 | − 491 |

F
| 773 | 925 | 834 | 802 | 889 |
|-----|-----|-----|-----|-----|
| − 148 | − 681 | − 417 | − 462 | − 285 |

**A**

$$461 - 267$$  $$889 - 803$$  $$473 - 295$$  $$815 - 167$$  $$337 - 302$$

**B**

$$854 - 280$$  $$526 - 377$$  $$909 - 114$$  $$800 - 281$$  $$823 - 662$$

**C**

$$953 - 733$$  $$919 - 362$$  $$907 - 286$$  $$870 - 790$$  $$600 - 380$$

**D**

$$834 - 623$$  $$963 - 227$$  $$520 - 516$$  $$722 - 531$$  $$904 - 212$$

**E**

$$398 - 180$$  $$779 - 617$$  $$778 - 684$$  $$868 - 238$$  $$561 - 508$$

**F**

$$381 - 185$$  $$926 - 252$$  $$812 - 610$$  $$933 - 738$$  $$906 - 290$$

A

$$820 - 772$$   $$797 - 285$$   $$532 - 463$$   $$657 - 186$$   $$930 - 169$$

B

$$746 - 107$$   $$394 - 219$$   $$509 - 428$$   $$836 - 216$$   $$485 - 350$$

C

$$869 - 457$$   $$780 - 389$$   $$881 - 758$$   $$846 - 441$$   $$844 - 631$$

D

$$548 - 473$$   $$215 - 182$$   $$741 - 592$$   $$447 - 126$$   $$840 - 649$$

E

$$705 - 364$$   $$951 - 826$$   $$752 - 726$$   $$367 - 316$$   $$666 - 104$$

F

$$971 - 793$$   $$615 - 210$$   $$785 - 505$$   $$729 - 361$$   $$138 - 135$$

**A**

| 434 | 479 | 906 | 906 | 465 |
| --- | --- | --- | --- | --- |
| − 185 | − 293 | − 516 | − 417 | − 116 |

**B**

| 892 | 759 | 849 | 323 | 996 |
| --- | --- | --- | --- | --- |
| − 296 | − 175 | − 119 | − 206 | − 974 |

**C**

| 549 | 950 | 780 | 939 | 845 |
| --- | --- | --- | --- | --- |
| − 320 | − 277 | − 742 | − 723 | − 279 |

**D**

| 336 | 497 | 997 | 501 | 737 |
| --- | --- | --- | --- | --- |
| − 305 | − 193 | − 666 | − 428 | − 203 |

**E**

| 304 | 775 | 482 | 402 | 578 |
| --- | --- | --- | --- | --- |
| − 103 | − 471 | − 384 | − 385 | − 339 |

**F**

| 573 | 636 | 773 | 788 | 918 |
| --- | --- | --- | --- | --- |
| − 425 | − 268 | − 769 | − 685 | − 165 |

**A**

$$872 - 513$$     $$946 - 541$$     $$958 - 479$$     $$761 - 443$$     $$865 - 774$$

**B**

$$775 - 331$$     $$890 - 150$$     $$793 - 439$$     $$845 - 101$$     $$426 - 404$$

**C**

$$490 - 169$$     $$671 - 660$$     $$861 - 309$$     $$797 - 334$$     $$341 - 152$$

**D**

$$898 - 142$$     $$686 - 174$$     $$928 - 435$$     $$419 - 284$$     $$984 - 522$$

**E**

$$954 - 555$$     $$376 - 130$$     $$751 - 372$$     $$325 - 215$$     $$726 - 132$$

**F**

$$918 - 626$$     $$452 - 213$$     $$765 - 621$$     $$623 - 220$$     $$758 - 358$$

**A**

| 884 | 600 | 914 | 927 | 744 |
|-----|-----|-----|-----|-----|
| − 790 | − 192 | − 521 | − 236 | − 547 |

**B**

| 562 | 717 | 913 | 586 | 823 |
|-----|-----|-----|-----|-----|
| − 446 | − 602 | − 853 | − 172 | − 570 |

**C**

| 870 | 518 | 895 | 722 | 512 |
|-----|-----|-----|-----|-----|
| − 253 | − 144 | − 214 | − 151 | − 412 |

**D**

| 968 | 581 | 972 | 701 | 580 |
|-----|-----|-----|-----|-----|
| − 606 | − 174 | − 921 | − 654 | − 540 |

**E**

| 757 | 839 | 939 | 715 | 773 |
|-----|-----|-----|-----|-----|
| − 749 | − 136 | − 503 | − 136 | − 147 |

**F**

| 797 | 368 | 755 | 903 | 987 |
|-----|-----|-----|-----|-----|
| − 757 | − 366 | − 738 | − 360 | − 346 |

**A**

966 − 278    926 − 454    301 − 254    902 − 585    692 − 687

**B**

943 − 461    809 − 371    819 − 148    697 − 510    288 − 243

**C**

697 − 512    935 − 749    668 − 569    499 − 487    518 − 324

**D**

654 − 374    392 − 176    661 − 321    328 − 195    360 − 103

**E**

667 − 168    228 − 141    967 − 650    570 − 365    709 − 470

**F**

522 − 246    917 − 839    696 − 635    196 − 150    495 − 188

A
| 700 | 398 | 894 | 850 | 616 |
| − 590 | − 142 | − 863 | − 660 | − 325 |

B
| 676 | 329 | 897 | 689 | 547 |
| − 276 | − 164 | − 594 | − 157 | − 133 |

C
| 733 | 338 | 777 | 700 | 889 |
| − 139 | − 140 | − 413 | − 298 | − 654 |

D
| 888 | 870 | 976 | 821 | 542 |
| − 329 | − 650 | − 571 | − 575 | − 336 |

E
| 471 | 995 | 628 | 770 | 888 |
| − 417 | − 775 | − 125 | − 312 | − 525 |

F
| 649 | 848 | 911 | 642 | 956 |
| − 503 | − 719 | − 207 | − 263 | − 502 |

A
```
   232        694        286        606        799
 - 222      - 262      - 269      - 509      - 362
```

B
```
   467        996        508        570        576
 - 275      - 447      - 307      - 143      - 539
```

C
```
   846        954        503        416        901
 - 761      - 447      - 292      - 118      - 828
```

D
```
   549        400        756        410        342
 - 123      - 137      - 732      - 197      - 313
```

E
```
   989        867        593        818        883
 - 267      - 403      - 341      - 392      - 242
```

F
```
   914        876        993        481        641
 - 142      - 605      - 395      - 288      - 403
```

A
$$\begin{array}{r} 185 \\ -\ 113 \\ \hline \end{array}$$
$$\begin{array}{r} 997 \\ -\ 311 \\ \hline \end{array}$$
$$\begin{array}{r} 538 \\ -\ 133 \\ \hline \end{array}$$
$$\begin{array}{r} 796 \\ -\ 280 \\ \hline \end{array}$$
$$\begin{array}{r} 352 \\ -\ 200 \\ \hline \end{array}$$

B
$$\begin{array}{r} 288 \\ -\ 274 \\ \hline \end{array}$$
$$\begin{array}{r} 776 \\ -\ 331 \\ \hline \end{array}$$
$$\begin{array}{r} 645 \\ -\ 240 \\ \hline \end{array}$$
$$\begin{array}{r} 653 \\ -\ 120 \\ \hline \end{array}$$
$$\begin{array}{r} 383 \\ -\ 114 \\ \hline \end{array}$$

C
$$\begin{array}{r} 902 \\ -\ 103 \\ \hline \end{array}$$
$$\begin{array}{r} 317 \\ -\ 186 \\ \hline \end{array}$$
$$\begin{array}{r} 723 \\ -\ 480 \\ \hline \end{array}$$
$$\begin{array}{r} 953 \\ -\ 652 \\ \hline \end{array}$$
$$\begin{array}{r} 991 \\ -\ 980 \\ \hline \end{array}$$

D
$$\begin{array}{r} 975 \\ -\ 437 \\ \hline \end{array}$$
$$\begin{array}{r} 907 \\ -\ 185 \\ \hline \end{array}$$
$$\begin{array}{r} 726 \\ -\ 119 \\ \hline \end{array}$$
$$\begin{array}{r} 441 \\ -\ 279 \\ \hline \end{array}$$
$$\begin{array}{r} 974 \\ -\ 573 \\ \hline \end{array}$$

E
$$\begin{array}{r} 869 \\ -\ 182 \\ \hline \end{array}$$
$$\begin{array}{r} 820 \\ -\ 643 \\ \hline \end{array}$$
$$\begin{array}{r} 816 \\ -\ 612 \\ \hline \end{array}$$
$$\begin{array}{r} 240 \\ -\ 190 \\ \hline \end{array}$$
$$\begin{array}{r} 772 \\ -\ 762 \\ \hline \end{array}$$

F
$$\begin{array}{r} 520 \\ -\ 168 \\ \hline \end{array}$$
$$\begin{array}{r} 537 \\ -\ 248 \\ \hline \end{array}$$
$$\begin{array}{r} 826 \\ -\ 675 \\ \hline \end{array}$$
$$\begin{array}{r} 707 \\ -\ 317 \\ \hline \end{array}$$
$$\begin{array}{r} 693 \\ -\ 623 \\ \hline \end{array}$$

A
```
  273        524        835        995        481
- 152      - 226      - 759      - 541      - 173
```

B
```
  851        632        840        916        988
- 649      - 222      - 728      - 695      - 184
```

C
```
  706        508        838        956        588
- 159      - 317      - 798      - 550      - 436
```

D
```
  789        806        720        405        513
- 306      - 585      - 478      - 174      - 472
```

E
```
  988        753        774        886        723
- 653      - 658      - 126      - 742      - 471
```

F
```
  384        678        794        596        959
- 358      - 206      - 209      - 329      - 832
```

**A**

```
  491        143        848        482        927
- 117      - 108      - 786      - 335      - 476
```

**B**

```
  856        828        691        891        557
- 459      - 547      - 654      - 103      - 242
```

**C**

```
  962        838        690        493        853
- 418      - 598      - 123      - 254      - 325
```

**D**

```
  721        546        855        502        798
- 614      - 402      - 730      - 301      - 284
```

**E**

```
  455        907        648        553        866
- 399      - 869      - 379      - 500      - 624
```

**F**

```
  952        953        805        871        947
- 824      - 341      - 265      - 187      - 398
```

A
$$917 - 402$$
$$405 - 330$$
$$792 - 321$$
$$575 - 545$$
$$486 - 170$$

B
$$490 - 371$$
$$736 - 417$$
$$777 - 122$$
$$381 - 276$$
$$743 - 650$$

C
$$785 - 345$$
$$927 - 146$$
$$726 - 711$$
$$588 - 318$$
$$597 - 210$$

D
$$681 - 119$$
$$913 - 258$$
$$970 - 571$$
$$788 - 253$$
$$441 - 232$$

E
$$947 - 552$$
$$432 - 346$$
$$766 - 568$$
$$655 - 106$$
$$492 - 235$$

F
$$349 - 104$$
$$935 - 507$$
$$457 - 435$$
$$914 - 239$$
$$481 - 323$$

A

$$868 - 725$$  $$528 - 253$$  $$527 - 308$$  $$901 - 163$$  $$566 - 458$$

B

$$410 - 160$$  $$688 - 608$$  $$818 - 133$$  $$627 - 100$$  $$928 - 467$$

C

$$639 - 332$$  $$527 - 290$$  $$905 - 464$$  $$879 - 319$$  $$977 - 970$$

D

$$296 - 143$$  $$860 - 710$$  $$964 - 381$$  $$752 - 196$$  $$749 - 592$$

E

$$203 - 122$$  $$972 - 476$$  $$557 - 441$$  $$990 - 378$$  $$882 - 207$$

F

$$575 - 458$$  $$960 - 871$$  $$532 - 206$$  $$459 - 365$$  $$638 - 349$$

A
$$935 - 732$$
$$635 - 485$$
$$576 - 411$$
$$834 - 285$$
$$917 - 704$$

B
$$651 - 484$$
$$856 - 739$$
$$519 - 505$$
$$687 - 390$$
$$890 - 506$$

C
$$441 - 120$$
$$957 - 580$$
$$894 - 446$$
$$304 - 260$$
$$823 - 609$$

D
$$981 - 533$$
$$995 - 895$$
$$751 - 182$$
$$804 - 698$$
$$852 - 663$$

E
$$594 - 262$$
$$719 - 593$$
$$474 - 176$$
$$340 - 281$$
$$592 - 523$$

F
$$128 - 108$$
$$919 - 340$$
$$950 - 582$$
$$473 - 469$$
$$821 - 414$$

**A**

| 713 | 583 | 780 | 201 | 359 |
|-----|-----|-----|-----|-----|
| − 705 | − 241 | − 377 | − 140 | − 245 |

**B**

| 419 | 995 | 962 | 663 | 772 |
|-----|-----|-----|-----|-----|
| − 117 | − 230 | − 101 | − 404 | − 271 |

**C**

| 484 | 989 | 518 | 710 | 691 |
|-----|-----|-----|-----|-----|
| − 121 | − 497 | − 312 | − 488 | − 306 |

**D**

| 444 | 495 | 896 | 542 | 293 |
|-----|-----|-----|-----|-----|
| − 443 | − 477 | − 845 | − 261 | − 129 |

**E**

| 803 | 566 | 899 | 763 | 844 |
|-----|-----|-----|-----|-----|
| − 254 | − 235 | − 634 | − 723 | − 612 |

**F**

| 649 | 926 | 862 | 698 | 529 |
|-----|-----|-----|-----|-----|
| − 479 | − 401 | − 518 | − 371 | − 323 |

A
| 895 | 645 | 746 | 817 | 781 |
| − 207 | − 146 | − 452 | − 688 | − 282 |

B
| 731 | 450 | 796 | 634 | 807 |
| − 504 | − 301 | − 180 | − 219 | − 404 |

C
| 635 | 734 | 914 | 528 | 361 |
| − 343 | − 642 | − 485 | − 374 | − 302 |

D
| 961 | 775 | 919 | 823 | 964 |
| − 157 | − 617 | − 828 | − 117 | − 796 |

E
| 431 | 994 | 332 | 477 | 429 |
| − 211 | − 724 | − 325 | − 279 | − 292 |

F
| 323 | 508 | 959 | 530 | 644 |
| − 177 | − 156 | − 767 | − 483 | − 456 |

Day: 30
Date:
Score: /30
Name:
Time:       :
Rating: ☆☆☆☆☆

A
$$907 - 257$$
$$457 - 404$$
$$899 - 631$$
$$891 - 217$$
$$768 - 529$$

B
$$807 - 230$$
$$813 - 419$$
$$540 - 138$$
$$576 - 379$$
$$736 - 715$$

C
$$797 - 411$$
$$800 - 140$$
$$708 - 282$$
$$869 - 147$$
$$268 - 120$$

D
$$975 - 828$$
$$740 - 250$$
$$954 - 251$$
$$492 - 434$$
$$962 - 603$$

E
$$371 - 331$$
$$710 - 278$$
$$812 - 336$$
$$771 - 624$$
$$578 - 153$$

F
$$913 - 512$$
$$934 - 814$$
$$295 - 135$$
$$948 - 350$$
$$796 - 171$$

A
| 762 − 576 | 697 − 389 | 840 − 732 | 637 − 328 | 234 − 229 |

B
| 871 − 424 | 443 − 104 | 831 − 812 | 810 − 526 | 413 − 322 |

C
| 874 − 794 | 786 − 324 | 496 − 439 | 795 − 197 | 866 − 721 |

D
| 202 − 180 | 911 − 683 | 298 − 269 | 499 − 329 | 495 − 370 |

E
| 271 − 193 | 828 − 741 | 749 − 742 | 953 − 115 | 670 − 286 |

F
| 412 − 120 | 668 − 473 | 940 − 557 | 348 − 191 | 349 − 212 |

Day: 32

Name:

Date:

Time:          :

Score:          /30

Rating: ☆☆☆☆☆

A
$$730 - 174$$
$$622 - 200$$
$$909 - 880$$
$$789 - 719$$
$$651 - 347$$

B
$$614 - 103$$
$$840 - 623$$
$$650 - 447$$
$$747 - 699$$
$$867 - 555$$

C
$$284 - 142$$
$$719 - 215$$
$$682 - 341$$
$$663 - 477$$
$$698 - 224$$

D
$$878 - 249$$
$$640 - 437$$
$$289 - 116$$
$$462 - 149$$
$$300 - 295$$

E
$$932 - 725$$
$$409 - 381$$
$$617 - 342$$
$$848 - 299$$
$$509 - 236$$

F
$$980 - 623$$
$$638 - 612$$
$$643 - 343$$
$$886 - 218$$
$$145 - 111$$

A
```
   829        965        886        941        629
 - 148      - 838      - 706      - 478      - 602
```

B
```
   839        471        903        407        905
 - 690      - 139      - 307      - 153      - 438
```

C
```
   978        849        998        783        434
 - 241      - 287      - 801      - 556      - 386
```

D
```
   522        613        989        960        616
 - 359      - 209      - 557      - 939      - 355
```

E
```
   693        918        659        730        875
 - 361      - 584      - 320      - 255      - 163
```

F
```
   900        738        489        609        935
 - 370      - 530      - 276      - 191      - 550
```

A

| 766 | 619 | 932 | 413 | 835 |
|---|---|---|---|---|
| − 240 | − 465 | − 469 | − 349 | − 188 |

B

| 709 | 643 | 737 | 559 | 641 |
|---|---|---|---|---|
| − 412 | − 327 | − 184 | − 374 | − 285 |

C

| 465 | 791 | 715 | 967 | 540 |
|---|---|---|---|---|
| − 118 | − 724 | − 613 | − 774 | − 173 |

D

| 983 | 736 | 351 | 221 | 278 |
|---|---|---|---|---|
| − 849 | − 642 | − 169 | − 172 | − 168 |

E

| 249 | 821 | 665 | 279 | 899 |
|---|---|---|---|---|
| − 186 | − 173 | − 109 | − 152 | − 781 |

F

| 586 | 976 | 939 | 832 | 407 |
|---|---|---|---|---|
| − 494 | − 462 | − 883 | − 270 | − 203 |

A

| 598 | 890 | 714 | 999 | 951 |
| - 500 | - 397 | - 675 | - 577 | - 609 |

B

| 607 | 836 | 494 | 719 | 749 |
| - 475 | - 661 | - 315 | - 217 | - 643 |

C

| 997 | 867 | 912 | 729 | 568 |
| - 515 | - 358 | - 710 | - 419 | - 123 |

D

| 323 | 435 | 995 | 978 | 565 |
| - 283 | - 340 | - 552 | - 338 | - 178 |

E

| 932 | 809 | 788 | 659 | 517 |
| - 687 | - 635 | - 399 | - 316 | - 108 |

F

| 484 | 904 | 737 | 597 | 382 |
| - 220 | - 634 | - 307 | - 511 | - 257 |

A

| 695 | 261 | 605 | 819 | 843 |
|-----|-----|-----|-----|-----|
| − 264 | − 127 | − 478 | − 496 | − 766 |

B

| 773 | 301 | 949 | 843 | 886 |
|-----|-----|-----|-----|-----|
| − 114 | − 108 | − 330 | − 566 | − 234 |

C

| 389 | 962 | 697 | 345 | 993 |
|-----|-----|-----|-----|-----|
| − 323 | − 817 | − 549 | − 168 | − 874 |

D

| 672 | 754 | 813 | 875 | 803 |
|-----|-----|-----|-----|-----|
| − 197 | − 323 | − 695 | − 264 | − 780 |

E

| 922 | 964 | 891 | 798 | 834 |
|-----|-----|-----|-----|-----|
| − 779 | − 326 | − 495 | − 390 | − 494 |

F

| 982 | 425 | 769 | 754 | 454 |
|-----|-----|-----|-----|-----|
| − 123 | − 284 | − 435 | − 379 | − 236 |

**A**

$$753 - 305$$   $$859 - 171$$   $$561 - 380$$   $$895 - 717$$   $$491 - 471$$

**B**

$$823 - 134$$   $$608 - 145$$   $$579 - 137$$   $$642 - 341$$   $$497 - 157$$

**C**

$$850 - 642$$   $$926 - 479$$   $$676 - 364$$   $$846 - 607$$   $$873 - 679$$

**D**

$$946 - 728$$   $$595 - 525$$   $$809 - 260$$   $$621 - 590$$   $$836 - 431$$

**E**

$$531 - 126$$   $$722 - 114$$   $$796 - 411$$   $$617 - 211$$   $$915 - 588$$

**F**

$$200 - 179$$   $$830 - 631$$   $$962 - 107$$   $$849 - 142$$   $$967 - 518$$

A
$$382 - 302$$   $$504 - 129$$   $$212 - 153$$   $$921 - 265$$   $$301 - 218$$

B
$$858 - 457$$   $$528 - 241$$   $$690 - 632$$   $$739 - 259$$   $$288 - 226$$

C
$$924 - 440$$   $$779 - 319$$   $$960 - 372$$   $$774 - 481$$   $$819 - 554$$

D
$$560 - 477$$   $$780 - 777$$   $$788 - 576$$   $$732 - 544$$   $$880 - 492$$

E
$$674 - 621$$   $$377 - 360$$   $$125 - 105$$   $$949 - 334$$   $$916 - 361$$

F
$$870 - 568$$   $$800 - 269$$   $$992 - 581$$   $$974 - 818$$   $$938 - 670$$

A
```
  772        959        772        780        639
- 745      - 715      - 383      - 265      - 284
```

B
```
  983        354        768        914        615
- 480      - 149      - 639      - 288      - 152
```

C
```
  968        479        626        929        856
- 521      - 319      - 191      - 406      - 750
```

D
```
  645        425        732        355        975
- 366      - 120      - 167      - 168      - 937
```

E
```
  499        909        782        668        901
- 330      - 176      - 709      - 117      - 824
```

F
```
  876        369        610        949        788
- 381      - 210      - 566      - 123      - 324
```

**A**

| 806 | 508 | 336 | 827 | 682 |
| − 303 | − 234 | − 213 | − 207 | − 436 |

**B**

| 302 | 941 | 667 | 527 | 908 |
| − 147 | − 893 | − 243 | − 217 | − 784 |

**C**

| 187 | 507 | 645 | 617 | 544 |
| − 172 | − 373 | − 593 | − 235 | − 122 |

**D**

| 737 | 658 | 165 | 278 | 531 |
| − 279 | − 649 | − 137 | − 151 | − 294 |

**E**

| 860 | 859 | 879 | 418 | 924 |
| − 292 | − 124 | − 219 | − 233 | − 410 |

**F**

| 899 | 849 | 802 | 663 | 837 |
| − 779 | − 389 | − 168 | − 450 | − 538 |

A
```
  896       997       295       578       603
- 132     - 529     - 138     - 387     - 527
```

B
```
  442       558       784       571       699
- 270     - 264     - 187     - 361     - 365
```

C
```
  595       815       983       938       670
- 563     - 565     - 493     - 743     - 621
```

D
```
  660       862       716       864       705
- 445     - 724     - 231     - 655     - 364
```

E
```
  902       575       951       921       574
- 863     - 566     - 479     - 745     - 392
```

F
```
  930       749       356       877       745
- 503     - 499     - 147     - 333     - 308
```

A
$$
\begin{array}{r} 924 \\ -\ 288 \\ \hline \end{array}
\qquad
\begin{array}{r} 376 \\ -\ 187 \\ \hline \end{array}
\qquad
\begin{array}{r} 952 \\ -\ 672 \\ \hline \end{array}
\qquad
\begin{array}{r} 779 \\ -\ 497 \\ \hline \end{array}
\qquad
\begin{array}{r} 684 \\ -\ 589 \\ \hline \end{array}
$$

B
$$
\begin{array}{r} 412 \\ -\ 238 \\ \hline \end{array}
\qquad
\begin{array}{r} 955 \\ -\ 194 \\ \hline \end{array}
\qquad
\begin{array}{r} 873 \\ -\ 500 \\ \hline \end{array}
\qquad
\begin{array}{r} 619 \\ -\ 236 \\ \hline \end{array}
\qquad
\begin{array}{r} 630 \\ -\ 186 \\ \hline \end{array}
$$

C
$$
\begin{array}{r} 753 \\ -\ 491 \\ \hline \end{array}
\qquad
\begin{array}{r} 796 \\ -\ 385 \\ \hline \end{array}
\qquad
\begin{array}{r} 533 \\ -\ 329 \\ \hline \end{array}
\qquad
\begin{array}{r} 723 \\ -\ 153 \\ \hline \end{array}
\qquad
\begin{array}{r} 359 \\ -\ 239 \\ \hline \end{array}
$$

D
$$
\begin{array}{r} 740 \\ -\ 737 \\ \hline \end{array}
\qquad
\begin{array}{r} 636 \\ -\ 320 \\ \hline \end{array}
\qquad
\begin{array}{r} 795 \\ -\ 283 \\ \hline \end{array}
\qquad
\begin{array}{r} 944 \\ -\ 204 \\ \hline \end{array}
\qquad
\begin{array}{r} 666 \\ -\ 622 \\ \hline \end{array}
$$

E
$$
\begin{array}{r} 612 \\ -\ 289 \\ \hline \end{array}
\qquad
\begin{array}{r} 705 \\ -\ 166 \\ \hline \end{array}
\qquad
\begin{array}{r} 863 \\ -\ 547 \\ \hline \end{array}
\qquad
\begin{array}{r} 243 \\ -\ 165 \\ \hline \end{array}
\qquad
\begin{array}{r} 879 \\ -\ 388 \\ \hline \end{array}
$$

F
$$
\begin{array}{r} 671 \\ -\ 221 \\ \hline \end{array}
\qquad
\begin{array}{r} 932 \\ -\ 442 \\ \hline \end{array}
\qquad
\begin{array}{r} 717 \\ -\ 626 \\ \hline \end{array}
\qquad
\begin{array}{r} 986 \\ -\ 409 \\ \hline \end{array}
\qquad
\begin{array}{r} 742 \\ -\ 171 \\ \hline \end{array}
$$

A
| 880 | 800 | 690 | 665 | 895 |
| − 605 | − 407 | − 465 | − 591 | − 720 |

B
| 427 | 559 | 706 | 175 | 796 |
| − 388 | − 520 | − 362 | − 126 | − 724 |

C
| 825 | 845 | 956 | 910 | 901 |
| − 676 | − 557 | − 330 | − 410 | − 873 |

D
| 882 | 940 | 604 | 877 | 467 |
| − 370 | − 501 | − 117 | − 315 | − 215 |

E
| 704 | 978 | 642 | 572 | 863 |
| − 692 | − 140 | − 478 | − 452 | − 434 |

F
| 694 | 844 | 308 | 578 | 757 |
| − 170 | − 774 | − 205 | − 209 | − 195 |

**A**

| 741 | 695 | 660 | 468 | 452 |
|-----|-----|-----|-----|-----|
| − 338 | − 463 | − 169 | − 303 | − 375 |

**B**

| 390 | 856 | 967 | 719 | 822 |
|-----|-----|-----|-----|-----|
| − 181 | − 741 | − 715 | − 618 | − 546 |

**C**

| 250 | 638 | 322 | 864 | 958 |
|-----|-----|-----|-----|-----|
| − 199 | − 267 | − 259 | − 675 | − 617 |

**D**

| 422 | 861 | 795 | 684 | 968 |
|-----|-----|-----|-----|-----|
| − 403 | − 280 | − 374 | − 363 | − 936 |

**E**

| 681 | 838 | 682 | 715 | 941 |
|-----|-----|-----|-----|-----|
| − 249 | − 391 | − 115 | − 300 | − 811 |

**F**

| 184 | 765 | 794 | 423 | 797 |
|-----|-----|-----|-----|-----|
| − 179 | − 668 | − 487 | − 281 | − 256 |

**A**

| 688 | 518 | 842 | 996 | 600 |
|---|---|---|---|---|
| − 688 | − 287 | − 273 | − 393 | − 139 |

**B**

| 535 | 455 | 753 | 775 | 767 |
|---|---|---|---|---|
| − 503 | − 340 | − 403 | − 506 | − 210 |

**C**

| 403 | 899 | 960 | 868 | 739 |
|---|---|---|---|---|
| − 355 | − 613 | − 742 | − 858 | − 252 |

**D**

| 487 | 673 | 885 | 974 | 955 |
|---|---|---|---|---|
| − 297 | − 541 | − 189 | − 571 | − 618 |

**E**

| 804 | 494 | 896 | 865 | 210 |
|---|---|---|---|---|
| − 792 | − 188 | − 570 | − 161 | − 133 |

**F**

| 705 | 346 | 541 | 421 | 853 |
|---|---|---|---|---|
| − 584 | − 206 | − 448 | − 396 | − 285 |

30

60

A
$$319 - 252$$   $$915 - 816$$   $$778 - 357$$   $$570 - 294$$   $$680 - 400$$

B
$$836 - 531$$   $$801 - 139$$   $$707 - 614$$   $$745 - 169$$   $$978 - 658$$

C
$$378 - 269$$   $$826 - 744$$   $$610 - 437$$   $$683 - 441$$   $$831 - 259$$

D
$$862 - 246$$   $$602 - 207$$   $$931 - 467$$   $$823 - 419$$   $$779 - 291$$

E
$$787 - 143$$   $$697 - 542$$   $$789 - 220$$   $$483 - 183$$   $$746 - 551$$

F
$$941 - 923$$   $$908 - 264$$   $$856 - 367$$   $$386 - 341$$   $$944 - 739$$

**A**

| | | | | |
|---|---|---|---|---|
| 496 − 204 | 475 − 378 | 328 − 119 | 768 − 102 | 623 − 201 |

**B**

| | | | | |
|---|---|---|---|---|
| 888 − 378 | 863 − 514 | 656 − 470 | 502 − 335 | 464 − 463 |

**C**

| | | | | |
|---|---|---|---|---|
| 916 − 376 | 485 − 247 | 912 − 849 | 643 − 638 | 336 − 243 |

**D**

| | | | | |
|---|---|---|---|---|
| 720 − 569 | 486 − 134 | 209 − 143 | 675 − 640 | 365 − 110 |

**E**

| | | | | |
|---|---|---|---|---|
| 497 − 204 | 967 − 489 | 817 − 132 | 663 − 543 | 722 − 650 |

**F**

| | | | | |
|---|---|---|---|---|
| 792 − 323 | 551 − 423 | 730 − 544 | 655 − 433 | 457 − 276 |

A
$$877 - 637$$
$$818 - 241$$
$$882 - 695$$
$$387 - 173$$
$$875 - 723$$

B
$$491 - 165$$
$$748 - 351$$
$$772 - 391$$
$$943 - 131$$
$$621 - 213$$

C
$$982 - 925$$
$$845 - 454$$
$$762 - 598$$
$$942 - 220$$
$$576 - 180$$

D
$$592 - 538$$
$$862 - 827$$
$$907 - 227$$
$$784 - 311$$
$$841 - 431$$

E
$$731 - 104$$
$$284 - 183$$
$$832 - 747$$
$$801 - 787$$
$$620 - 414$$

F
$$698 - 234$$
$$716 - 344$$
$$698 - 311$$
$$930 - 450$$
$$660 - 608$$

A
```
  970        665        373        260        641
- 252      - 498      - 228      - 230      - 518
```

B
```
  884        878        406        866        584
- 833      - 486      - 260      - 830      - 537
```

C
```
  699        745        804        790        577
- 547      - 334      - 676      - 493      - 264
```

D
```
  367        718        955        559        972
- 232      - 425      - 640      - 100      - 886
```

E
```
  646        591        571        601        757
- 166      - 173      - 193      - 101      - 296
```

F
```
  863        804        877        462        529
- 516      - 771      - 620      - 300      - 490
```

A
```
   880        770        352        905        276
 - 287      - 204      - 198      - 622      - 267
```

B
```
   654        911        750        610        920
 - 107      - 133      - 630      - 208      - 338
```

C
```
   650        912        678        996        713
 - 453      - 866      - 280      - 160      - 219
```

D
```
   806        719        864        496        609
 - 789      - 609      - 247      - 395      - 381
```

E
```
   380        897        986        602        687
 - 167      - 695      - 309      - 299      - 355
```

F
```
   303        882        632        400        274
 - 282      - 679      - 102      - 374      - 185
```

**A**

| 826 | 774 | 859 | 362 | 907 |
|-----|-----|-----|-----|-----|
| − 277 | − 177 | − 557 | − 185 | − 321 |

**B**

| 328 | 848 | 895 | 326 | 873 |
|-----|-----|-----|-----|-----|
| − 102 | − 790 | − 351 | − 288 | − 234 |

**C**

| 909 | 869 | 837 | 819 | 883 |
|-----|-----|-----|-----|-----|
| − 148 | − 177 | − 462 | − 728 | − 331 |

**D**

| 900 | 659 | 628 | 961 | 775 |
|-----|-----|-----|-----|-----|
| − 114 | − 233 | − 303 | − 856 | − 254 |

**E**

| 636 | 936 | 742 | 676 | 528 |
|-----|-----|-----|-----|-----|
| − 145 | − 311 | − 568 | − 218 | − 512 |

**F**

| 824 | 503 | 471 | 693 | 553 |
|-----|-----|-----|-----|-----|
| − 372 | − 110 | − 440 | − 668 | − 397 |

# Addition Answer Key Sheet

| Day | | | | | | | | | | | | | | | | | | |
|---|---|---|---|---|---|---|---|---|---|---|---|---|---|---|---|---|---|---|
| DAY 1 | A | 1181 | 1596 | 840 | 1552 | 1418 | B | 1005 | 1098 | 1655 | 722 | 1440 | C | 1116 | 778 | 743 | 966 | 516 |
| | D | 601 | 1595 | 805 | 850 | 905 | E | 1173 | 1337 | 1268 | 1106 | 1181 | F | 1632 | 1437 | 1152 | 1379 | 1469 |
| DAY 2 | A | 352 | 1189 | 1732 | 1132 | 725 | B | 1063 | 1019 | 1972 | 1787 | 928 | C | 898 | 1601 | 946 | 353 | 1513 |
| | D | 1431 | 717 | 1546 | 1142 | 828 | E | 1355 | 1074 | 657 | 970 | 775 | F | 1107 | 1154 | 1167 | 535 | 1130 |
| DAY 3 | A | 1156 | 1089 | 1701 | 656 | 1128 | B | 380 | 816 | 1009 | 1005 | 1031 | C | 1458 | 1882 | 805 | 713 | 963 |
| | D | 673 | 604 | 1255 | 1377 | 1746 | E | 827 | 1444 | 737 | 1285 | 1373 | F | 1013 | 529 | 1235 | 877 | 1271 |
| DAY 4 | A | 1106 | 334 | 1253 | 572 | 1631 | B | 1105 | 1727 | 1476 | 1336 | 955 | C | 1680 | 1079 | 946 | 1004 | 1194 |
| | D | 1711 | 1418 | 595 | 1275 | 1128 | E | 1401 | 1189 | 1400 | 1608 | 1058 | F | 1456 | 1136 | 1488 | 1563 | 758 |
| DAY 5 | A | 1060 | 1699 | 1359 | 537 | 1635 | B | 884 | 603 | 1335 | 514 | 1600 | C | 1120 | 1170 | 344 | 850 | 1244 |
| | D | 1197 | 1326 | 1326 | 1156 | 931 | E | 969 | 1505 | 1163 | 1488 | 689 | F | 1050 | 1489 | 1064 | 1198 | 1345 |
| DAY 6 | A | 1252 | 615 | 1495 | 1700 | 1359 | B | 1658 | 721 | 1293 | 1047 | 851 | C | 1008 | 1024 | 1437 | 1478 | 453 |
| | D | 1824 | 1321 | 1437 | 1095 | 902 | E | 731 | 641 | 1169 | 556 | 1275 | F | 1489 | 1566 | 789 | 910 | 1102 |
| DAY 7 | A | 1240 | 1237 | 1108 | 1309 | 936 | B | 543 | 1542 | 972 | 835 | 705 | C | 939 | 985 | 856 | 1392 | 1434 |
| | D | 1349 | 1038 | 604 | 1619 | 778 | E | 847 | 717 | 994 | 639 | 562 | F | 1478 | 1116 | 995 | 1030 | 645 |
| DAY 8 | A | 945 | 1352 | 1944 | 489 | 1047 | B | 1091 | 967 | 468 | 674 | 1053 | C | 1633 | 1260 | 480 | 1072 | 691 |
| | D | 1461 | 886 | 638 | 446 | 800 | E | 1400 | 923 | 1115 | 540 | 1330 | F | 1853 | 1267 | 1578 | 1126 | 755 |
| DAY 9 | A | 1220 | 768 | 1331 | 1245 | 883 | B | 1893 | 1090 | 1192 | 1010 | 589 | C | 1731 | 1001 | 1483 | 1022 | 1567 |
| | D | 1026 | 1870 | 672 | 987 | 1400 | E | 536 | 1085 | 1921 | 461 | 1619 | F | 474 | 1478 | 486 | 510 | 795 |
| DAY 10 | A | 1472 | 1079 | 1722 | 1557 | 1192 | B | 450 | 1203 | 1220 | 1228 | 319 | C | 947 | 1594 | 1356 | 1043 | 1203 |
| | D | 1126 | 1484 | 1051 | 1506 | 695 | E | 1136 | 1752 | 1058 | 914 | 1367 | F | 1108 | 833 | 991 | 1066 | 1264 |
| DAY 11 | A | 693 | 1445 | 772 | 1153 | 1250 | B | 1925 | 568 | 864 | 1093 | 1231 | C | 1504 | 996 | 1861 | 564 | 1205 |
| | D | 1036 | 987 | 729 | 928 | 1398 | E | 755 | 740 | 1460 | 640 | 1037 | F | 460 | 1428 | 1292 | 700 | 752 |
| DAY 12 | A | 1176 | 1120 | 830 | 1640 | 1223 | B | 961 | 1120 | 822 | 1066 | 1238 | C | 1427 | 1001 | 510 | 1779 | 966 |
| | D | 1658 | 562 | 1292 | 1373 | 1135 | E | 1415 | 577 | 1216 | 937 | 948 | F | 1025 | 943 | 896 | 945 | 1268 |
| DAY 13 | A | 876 | 661 | 849 | 1048 | 1008 | B | 1521 | 1531 | 1939 | 732 | 900 | C | 816 | 768 | 542 | 587 | 1640 |
| | D | 673 | 563 | 1196 | 318 | 1394 | E | 907 | 1311 | 668 | 1083 | 1469 | F | 1196 | 779 | 699 | 903 | 1286 |
| DAY 14 | A | 1764 | 1185 | 960 | 930 | 1509 | B | 426 | 1592 | 548 | 1102 | 800 | C | 849 | 684 | 1761 | 683 | 925 |
| | D | 1896 | 1006 | 1274 | 1451 | 422 | E | 904 | 632 | 1081 | 1529 | 996 | F | 1353 | 1388 | 758 | 636 | 1258 |
| DAY 15 | A | 737 | 1466 | 1588 | 888 | 1653 | B | 959 | 1051 | 1047 | 771 | 918 | C | 786 | 1062 | 860 | 1249 | 1096 |
| | D | 1391 | 1575 | 515 | 456 | 1018 | E | 1215 | 1755 | 573 | 986 | 1182 | F | 1057 | 966 | 771 | 1107 | 1021 |
| DAY 16 | A | 723 | 910 | 1205 | 1065 | 907 | B | 721 | 1319 | 1186 | 935 | 866 | C | 899 | 984 | 1588 | 1465 | 542 |
| | D | 720 | 1470 | 768 | 687 | 429 | E | 1258 | 1487 | 1387 | 1102 | 686 | F | 432 | 765 | 1277 | 1243 | 480 |
| DAY 17 | A | 1642 | 1353 | 1559 | 815 | 1139 | B | 1197 | 705 | 1369 | 1258 | 1036 | C | 797 | 866 | 1019 | 1410 | 1815 |
| | D | 578 | 832 | 1129 | 637 | 1245 | E | 1069 | 1015 | 1788 | 577 | 586 | F | 1051 | 1131 | 1238 | 1165 | 1163 |
| DAY 18 | A | 872 | 1228 | 1128 | 903 | 849 | B | 973 | 1133 | 1236 | 1319 | 1106 | C | 692 | 842 | 860 | 1484 | 1127 |
| | D | 1181 | 1042 | 1217 | 433 | 829 | E | 1314 | 1175 | 1115 | 1070 | 634 | F | 255 | 576 | 752 | 653 | 1213 |
| DAY 19 | A | 842 | 1021 | 1190 | 1120 | 1056 | B | 1260 | 1030 | 1432 | 775 | 1294 | C | 1239 | 1444 | 1242 | 1524 | 1577 |
| | D | 784 | 1005 | 1500 | 1292 | 1573 | E | 943 | 1106 | 945 | 1441 | 833 | F | 1083 | 1469 | 1964 | 1514 | 1036 |
| DAY 20 | A | 1308 | 986 | 1089 | 920 | 1621 | B | 1917 | 833 | 1216 | 1845 | 1315 | C | 928 | 353 | 1770 | 1006 | 1256 |
| | D | 460 | 1416 | 854 | 1573 | 758 | E | 638 | 882 | 1597 | 1169 | 1171 | F | 661 | 988 | 877 | 838 | 1986 |
| DAY 21 | A | 985 | 678 | 966 | 1667 | 1032 | B | 813 | 1280 | 1249 | 1515 | 1009 | C | 528 | 457 | 1214 | 1101 | 1211 |
| | D | 1498 | 601 | 1346 | 1081 | 845 | E | 975 | 1186 | 1625 | 1373 | 577 | F | 820 | 1245 | 1266 | 1243 | 799 |
| DAY 22 | A | 1024 | 1932 | 1080 | 1650 | 1122 | B | 1169 | 1257 | 682 | 1483 | 836 | C | 1735 | 715 | 652 | 1612 | 936 |
| | D | 1365 | 993 | 1753 | 1554 | 542 | E | 393 | 1144 | 1609 | 1177 | 759 | F | 1311 | 680 | 627 | 984 | 1426 |
| DAY 23 | A | 1151 | 1174 | 1434 | 1098 | 1267 | B | 1099 | 1768 | 1520 | 1236 | 761 | C | 1351 | 1019 | 1102 | 423 | 352 |
| | D | 1458 | 764 | 1263 | 1293 | 1533 | E | 1752 | 1317 | 769 | 1410 | 963 | F | 785 | 967 | 1112 | 332 | 1224 |
| DAY 24 | A | 1281 | 1270 | 1356 | 1646 | 1309 | B | 1029 | 455 | 796 | 898 | 1156 | C | 740 | 949 | 1529 | 1401 | 703 |
| | D | 1874 | 1012 | 1150 | 276 | 1445 | E | 1387 | 1599 | 555 | 1865 | 1787 | F | 1863 | 907 | 892 | 1809 | 1161 |
| DAY 25 | A | 712 | 669 | 599 | 737 | 832 | B | 1375 | 878 | 1067 | 991 | 897 | C | 888 | 1368 | 1432 | 1208 | 1072 |
| | D | 851 | 848 | 597 | 1287 | 503 | E | 1429 | 1753 | 1376 | 823 | 1616 | F | 1001 | 1507 | 1094 | 1250 | 958 |
| DAY 26 | A | 521 | 1043 | 820 | 1313 | 1319 | B | 863 | 1008 | 805 | 1434 | 903 | C | 690 | 1743 | 1669 | 1280 | 553 |
| | D | 1430 | 885 | 1391 | 1742 | 374 | E | 1220 | 1450 | 838 | 1064 | 756 | F | 1531 | 1430 | 785 | 736 | 947 |
| DAY 27 | A | 1206 | 1636 | 1012 | 1191 | 1143 | B | 1405 | 540 | 1149 | 576 | 944 | C | 708 | 1380 | 771 | 536 | 1245 |
| | D | 1092 | 1122 | 745 | 751 | 1308 | E | 304 | 530 | 1095 | 1272 | 1310 | F | 1002 | 1456 | 881 | 929 | 738 |
| DAY 28 | A | 1151 | 435 | 1258 | 520 | 1352 | B | 1013 | 1401 | 1106 | 1403 | 867 | C | 1585 | 600 | 1022 | 1456 | 781 |
| | D | 1700 | 1441 | 1124 | 1313 | 1132 | E | 1133 | 574 | 916 | 1036 | 740 | F | 1148 | 1157 | 753 | 1340 | 1346 |
| DAY 29 | A | 791 | 808 | 1257 | 1043 | 734 | B | 632 | 1419 | 1195 | 805 | 1310 | C | 914 | 1125 | 1021 | 804 | 811 |
| | D | 1222 | 1011 | 1122 | 809 | 1735 | E | 1056 | 845 | 1231 | 1606 | 1132 | F | 1754 | 1049 | 853 | 1214 | 432 |
| DAY 30 | A | 1485 | 1350 | 881 | 1256 | 1085 | B | 1029 | 661 | 1358 | 1426 | 801 | C | 271 | 1375 | 1084 | 851 | 1062 |
| | D | 1297 | 1425 | 1167 | 740 | 1595 | E | 1102 | 552 | 722 | 986 | 741 | F | 1596 | 1093 | 1107 | 1452 | 779 |
| DAY 31 | A | 607 | 717 | 1223 | 1234 | 646 | B | 595 | 1271 | 1089 | 1428 | 761 | C | 1627 | 1544 | 1528 | 1016 | 1024 |
| | D | 751 | 1318 | 1468 | 1574 | 1087 | E | 1197 | 1783 | 784 | 1258 | 1802 | F | 1513 | 1271 | 1519 | 836 | 1804 |
| DAY 32 | A | 1185 | 1153 | 1133 | 1352 | 1558 | B | 1154 | 1229 | 1063 | 1370 | 614 | C | 658 | 1076 | 1443 | 1423 | 1617 |
| | D | 905 | 1267 | 941 | 622 | 1555 | E | 1177 | 1175 | 1135 | 1831 | 761 | F | 1074 | 675 | 1387 | 1225 | 1063 |
| DAY 33 | A | 1182 | 681 | 1210 | 1061 | 1349 | B | 542 | 1053 | 966 | 852 | 285 | C | 491 | 1468 | 1918 | 841 | 1180 |
| | D | 844 | 915 | 925 | 907 | 762 | E | 849 | 1324 | 1019 | 678 | 1486 | F | 447 | 1353 | 810 | 762 | 871 |
| DAY 34 | A | 982 | 1523 | 1352 | 1547 | 649 | B | 1053 | 1035 | 719 | 1538 | 1802 | C | 627 | 893 | 457 | 492 | 794 |
| | D | 251 | 1187 | 1384 | 1125 | 1171 | E | 965 | 1302 | 1716 | 1935 | 1324 | F | 914 | 910 | 1316 | 893 | 533 |
| DAY 35 | A | 1538 | 1359 | 466 | 1531 | 1159 | B | 512 | 1150 | 1783 | 1077 | 1902 | C | 1294 | 806 | 1513 | 1110 | 1568 |
| | D | 1661 | 1470 | 1328 | 1779 | 1417 | E | 760 | 1034 | 1341 | 1508 | 963 | F | 1437 | 1973 | 1613 | 1197 | 1361 |
| DAY 36 | A | 1043 | 710 | 1585 | 612 | 578 | B | 1110 | 624 | 1023 | 1833 | 1326 | C | 702 | 1596 | 985 | 924 | 1702 |
| | D | 771 | 1023 | 1434 | 861 | 1025 | E | 1554 | 535 | 1222 | 1036 | 1672 | F | 672 | 918 | 1355 | 1588 | 1374 |
| DAY 37 | A | 965 | 701 | 1264 | 1597 | 1327 | B | 1615 | 1629 | 1330 | 941 | 1395 | C | 585 | 1188 | 673 | 1087 | 331 |
| | D | 973 | 1323 | 1448 | 546 | 1001 | E | 987 | 931 | 834 | 1707 | 886 | F | 1768 | 965 | 1680 | 1562 | 1559 |
| DAY 38 | A | 1162 | 1219 | 1163 | 1164 | 1399 | B | 1740 | 673 | 1081 | 1855 | 888 | C | 473 | 909 | 1302 | 447 | 789 |
| | D | 1040 | 825 | 714 | 970 | 1165 | E | 952 | 1082 | 1326 | 931 | 1238 | F | 859 | 1349 | 1072 | 1226 | 1123 |
| DAY 39 | A | 856 | 840 | 1456 | 1646 | 1222 | B | 1214 | 416 | 1053 | 1306 | 1018 | C | 1319 | 1131 | 703 | 1062 | 1121 |
| | D | 995 | 1876 | 922 | 1176 | 801 | E | 800 | 1164 | 986 | 1537 | 809 | F | 576 | 1421 | 569 | 1202 | 1421 |
| DAY 40 | A | 427 | 1097 | 1338 | 1191 | 1331 | B | 1310 | 714 | 896 | 1082 | 1181 | C | 601 | 1403 | 771 | 712 | 495 |
| | D | 1510 | 642 | 1493 | 1063 | 1313 | E | 985 | 759 | 1329 | 1113 | 338 | F | 1730 | 500 | 1471 | 1903 | 1898 |
| DAY 41 | A | 1155 | 949 | 1428 | 1627 | 839 | B | 855 | 479 | 1203 | 1378 | 852 | C | 552 | 755 | 1165 | 725 | 1146 |
| | D | 1097 | 1241 | 1059 | 1125 | 1306 | E | 1145 | 932 | 994 | 547 | 1295 | F | 755 | 959 | 742 | 959 | 1217 |
| DAY 42 | A | 839 | 629 | 1642 | 918 | 1317 | B | 1277 | 1411 | 1642 | 532 | 1205 | C | 1803 | 767 | 1642 | 1469 | 1293 |
| | D | 1812 | 1360 | 807 | 1689 | 1446 | E | 1476 | 794 | 1153 | 1037 | 1067 | F | 1073 | 1090 | 1398 | 1026 | 1486 |
| DAY 43 | A | 774 | 1061 | 1423 | 1537 | 1389 | B | 581 | 1517 | 728 | 1203 | 1247 | C | 1588 | 1607 | 1527 | 1210 | 1048 |
| | D | 1081 | 1225 | 1055 | 338 | 1333 | E | 1466 | 1164 | 1103 | 1264 | 1241 | F | 1052 | 611 | 1258 | 1361 | 830 |
| DAY 44 | A | 908 | 1111 | 1421 | 1795 | 1078 | B | 1238 | 844 | 1730 | 1287 | 1010 | C | 1158 | 952 | 1041 | 1163 | 1445 |
| | D | 1564 | 1592 | 953 | 1483 | 1149 | E | 1544 | 700 | 722 | 1073 | 959 | F | 1051 | 580 | 1170 | 1212 | 1445 |
| DAY 45 | A | 1283 | 1395 | 1654 | 1525 | 1152 | B | 773 | 894 | 1141 | 1599 | 1040 | C | 1691 | 1546 | 1715 | 1044 | 965 |
| | D | 558 | 977 | 1073 | 1161 | 784 | E | 621 | 1086 | 1198 | 672 | 1428 | F | 814 | 1151 | 1143 | 897 | 804 |
| DAY 46 | A | 1096 | 1302 | 1518 | 1497 | 1935 | B | 1081 | 1778 | 1007 | 689 | 1250 | C | 1268 | 926 | 1198 | 1223 | 1516 |
| | D | 891 | 1286 | 787 | 979 | 528 | E | 1392 | 645 | 1067 | 659 | 721 | F | 520 | 1113 | 1534 | 965 | 408 |
| DAY 47 | A | 1443 | 1236 | 447 | 1420 | 1033 | B | 830 | 1057 | 1388 | 746 | 715 | C | 1441 | 1811 | 1347 | 1013 | 741 |
| | D | 1547 | 1086 | 1117 | 1030 | 516 | E | 839 | 1025 | 1823 | 1239 | 1181 | F | 389 | 730 | 854 | 1395 | 1015 |
| DAY 48 | A | 1220 | 1725 | 1943 | 952 | 1038 | B | 982 | 550 | 1400 | 1425 | 1743 | C | 727 | 790 | 1383 | 248 | 1298 |
| | D | 1681 | 335 | 992 | 603 | 1757 | E | 897 | 812 | 1561 | 737 | 721 | F | 1129 | 1161 | 511 | 921 | 940 |
| DAY 49 | A | 388 | 1366 | 1480 | 1674 | 335 | B | 1178 | 1776 | 1239 | 1235 | 1328 | C | 1583 | 1309 | 522 | 1589 | 1136 |
| | D | 674 | 883 | 1332 | 1506 | 1451 | E | 1030 | 1207 | 1271 | 1386 | 1035 | F | 1017 | 1347 | 1310 | 1608 | 332 |
| DAY 50 | A | 529 | 525 | 1508 | 1101 | 1101 | B | 1086 | 1418 | 1309 | 1046 | 1412 | C | 577 | 1492 | 1542 | 1213 | 1373 |
| | D | 623 | 926 | 1258 | 1181 | 1395 | E | 1282 | 1547 | 1091 | 449 | 781 | F | 757 | 1060 | 1499 | 1266 | 1411 |
| DAY 51 | A | 327 | 516 | 952 | 1014 | 500 | B | 858 | 1309 | 598 | 1328 | 976 | C | 1678 | 963 | 896 | 1211 | 1261 |
| | D | 1728 | 797 | 994 | 1598 | 804 | E | 1326 | 1291 | 911 | 868 | 1202 | F | 612 | 1089 | 525 | 1311 | 1290 |

Answer Key Mapping:

DAY 1 — A: 1181, 1596, 840, 1552, 1418 — B: 1005, 1098, 1655, 722, 1440
D: 601, 1595, 805, 850, 905 — E: 1173, 1337, 1268, 1106, 1181

Two rows (A-F) with five columns of solutions for each day.

Day 1 Problems & Solutions

A: 396 + 785 = 1181; 981 + 615 = 1596; 601 + 239 = 840; 795 + 757 = 1552; 647 + 771 = 1418

# Subtraction Answer Key Sheet

| DAY | Row | | | | | |
|---|---|---|---|---|---|---|
| 1 | A | 600, | 33, | 410, | 188, | 252 |
| 1 | D | 81, | 617, | 438, | 336, | 279 |
| 2 | A | 65, | 94, | 430, | 129, | 521 |
| 2 | D | 632, | 588, | 377, | 85, | 100 |
| 3 | A | 667, | 209, | 93, | 420, | 313 |
| 3 | D | 742, | 156, | 206, | 59, | 12 |
| 4 | A | 207, | 248, | 367, | 474, | 39 |
| 4 | D | 132, | 94, | 14, | 549, | 42 |
| 5 | A | 109, | 448, | 155, | 515, | 20 |
| 5 | D | 90, | 493, | 308, | 101, | 96 |
| 6 | A | 155, | 369, | 644, | 286, | 88 |
| 6 | D | 148, | 408, | 203, | 348, | 613 |
| 7 | A | 436, | 57, | 309, | 517, | 495 |
| 7 | D | 631, | 284, | 227, | 90, | 160 |
| 8 | A | 43, | 118, | 162, | 115, | 423 |
| 8 | D | 193, | 31, | 79, | 286, | 282 |
| 9 | A | 492, | 135, | 736, | 111, | 159 |
| 9 | D | 470, | 200, | 627, | 580, | 85 |
| 10 | A | 123, | 175, | 394, | 139, | 191 |
| 10 | D | 356, | 158, | 191, | 65, | 336 |
| 11 | A | 94, | 203, | 375, | 512, | 172 |
| 11 | D | 289, | 8, | 562, | 182, | 65 |
| 12 | A | 159, | 764, | 700, | 441, | 101 |
| 12 | D | 229, | 141, | 139, | 665, | 306 |
| 13 | A | 277, | 299, | 20, | 136, | 264 |
| 13 | D | 130, | 49, | 48, | 124, | 150 |
| 14 | A | 194, | 86, | 178, | 648, | 35 |
| 14 | D | 211, | 736, | 4, | 191, | 692 |
| 15 | A | 48, | 512, | 69, | 471, | 761 |
| 15 | D | 75, | 33, | 149, | 321, | 191 |
| 16 | A | 249, | 186, | 390, | 489, | 349 |
| 16 | D | 31, | 304, | 331, | 73, | 534 |
| 17 | A | 359, | 405, | 479, | 318, | 91 |
| 17 | D | 756, | 512, | 493, | 135, | 462 |
| 18 | A | 94, | 408, | 393, | 691, | 197 |
| 18 | D | 362, | 407, | 51, | 47, | 40 |
| 19 | A | 688, | 472, | 47, | 317, | 5 |
| 19 | D | 280, | 216, | 340, | 133, | 257 |
| 20 | A | 110, | 256, | 31, | 190, | 291 |
| 20 | D | 559, | 220, | 405, | 246, | 206 |
| 21 | A | 10, | 432, | 17, | 97, | 437 |
| 21 | D | 426, | 263, | 24, | 213, | 29 |
| 22 | A | 72, | 686, | 405, | 516, | 152 |
| 22 | D | 538, | 722, | 607, | 162, | 401 |
| 23 | A | 121, | 298, | 76, | 454, | 308 |
| 23 | D | 483, | 221, | 242, | 231, | 41 |
| 24 | A | 374, | 35, | 62, | 147, | 451 |
| 24 | D | 107, | 144, | 125, | 201, | 514 |
| 25 | A | 515, | 75, | 471, | 30, | 316 |
| 25 | D | 562, | 655, | 399, | 535, | 209 |
| 26 | A | 143, | 275, | 219, | 738, | 108 |
| 26 | D | 153, | 150, | 583, | 556, | 157 |
| 27 | A | 203, | 150, | 165, | 549, | 213 |
| 27 | D | 448, | 100, | 569, | 106, | 189 |
| 28 | A | 8, | 342, | 403, | 61, | 114 |
| 28 | D | 1, | 18, | 51, | 281, | 164 |
| 29 | A | 688, | 499, | 294, | 129, | 499 |
| 29 | D | 804, | 158, | 91, | 706, | 168 |
| 30 | A | 650, | 53, | 268, | 674, | 239 |
| 30 | D | 147, | 490, | 703, | 58, | 359 |
| 31 | A | 186, | 308, | 108, | 309, | 5 |
| 31 | D | 22, | 228, | 29, | 170, | 125 |
| 32 | A | 556, | 422, | 29, | 70, | 304 |
| 32 | D | 629, | 203, | 173, | 313, | 5 |
| 33 | A | 681, | 127, | 180, | 463, | 27 |
| 33 | D | 163, | 404, | 432, | 21, | 261 |
| 34 | A | 526, | 154, | 463, | 64, | 647 |
| 34 | D | 134, | 94, | 182, | 49, | 110 |
| 35 | A | 98, | 493, | 39, | 422, | 342 |
| 35 | D | 40, | 95, | 443, | 640, | 387 |
| 36 | A | 431, | 134, | 127, | 323, | 77 |
| 36 | D | 475, | 431, | 118, | 611, | 23 |
| 37 | A | 448, | 688, | 181, | 178, | 20 |
| 37 | D | 218, | 70, | 549, | 31, | 405 |
| 38 | A | 80, | 375, | 59, | 656, | 83 |
| 38 | D | 83, | 3, | 212, | 188, | 388 |
| 39 | A | 27, | 244, | 389, | 515, | 355 |
| 39 | D | 279, | 305, | 565, | 187, | 38 |
| 40 | A | 503, | 274, | 123, | 620, | 246 |
| 40 | D | 458, | 9, | 28, | 127, | 237 |
| 41 | A | 764, | 468, | 157, | 191, | 76 |
| 41 | D | 215, | 138, | 485, | 209, | 341 |
| 42 | A | 636, | 189, | 280, | 282, | 95 |
| 42 | D | 3, | 316, | 512, | 740, | 44 |
| 43 | A | 275, | 393, | 225, | 74, | 175 |
| 43 | D | 512, | 439, | 487, | 562, | 252 |
| 44 | A | 403, | 232, | 491, | 165, | 77 |
| 44 | D | 19, | 581, | 421, | 321, | 32 |
| 45 | A | 0, | 231, | 569, | 603, | 461 |
| 45 | D | 190, | 132, | 696, | 403, | 337 |
| 46 | A | 67, | 99, | 421, | 276, | 280 |
| 46 | D | 616, | 395, | 464, | 404, | 488 |
| 47 | A | 292, | 97, | 209, | 666, | 422 |
| 47 | D | 151, | 352, | 66, | 35, | 255 |
| 48 | A | 240, | 577, | 187, | 214, | 152 |
| 48 | D | 54, | 35, | 680, | 473, | 410 |
| 49 | A | 718, | 167, | 145, | 30, | 123 |
| 49 | D | 135, | 293, | 315, | 459, | 86 |
| 50 | A | 593, | 566, | 154, | 283, | 9 |
| 50 | D | 17, | 110, | 617, | 101, | 228 |
| 51 | A | 549, | 597, | 302, | 177, | 586 |
| 51 | D | 786, | 426, | 325, | 105, | 521 |

| DAY | Row | | | | | |
|---|---|---|---|---|---|---|
| 1 | B | 233, | 418, | 81, | 262, | 341 |
| 1 | E | 385, | 416, | 504, | 512, | 245 |
| 2 | B | 550, | 267, | 161, | 487, | 229 |
| 2 | E | 21, | 806, | 631, | 502, | 386 |
| 3 | B | 389, | 478, | 143, | 299, | 94 |
| 3 | E | 275, | 24, | 98, | 227, | 85 |
| 4 | B | 565, | 274, | 283, | 504, | 216 |
| 4 | E | 220, | 443, | 693, | 80, | 3 |
| 5 | B | 724, | 123, | 273, | 165, | 136 |
| 5 | E | 544, | 301, | 381, | 160, | 39 |
| 6 | B | 111, | 498, | 54, | 222, | 606 |
| 6 | E | 164, | 87, | 326, | 187, | 437 |
| 7 | B | 27, | 44, | 462, | 592, | 680 |
| 7 | E | 317, | 643, | 20, | 410, | 231 |
| 8 | B | 161, | 233, | 293, | 568, | 53 |
| 8 | E | 46, | 140, | 385, | 283, | 251 |
| 9 | B | 133, | 705, | 790, | 146, | 73 |
| 9 | E | 731, | 395, | 149, | 132, | 472 |
| 10 | B | 110, | 107, | 414, | 540, | 12 |
| 10 | E | 113, | 247, | 229, | 167, | 61 |
| 11 | B | 693, | 485, | 203, | 64, | 306 |
| 11 | E | 305, | 271, | 49, | 274, | 326 |
| 12 | B | 332, | 37, | 92, | 417, | 195 |
| 12 | E | 167, | 334, | 285, | 367, | 120 |
| 13 | B | 506, | 275, | 287, | 502, | 30 |
| 13 | E | 411, | 320, | 3, | 1, | 414 |
| 14 | B | 574, | 149, | 795, | 519, | 161 |
| 14 | E | 218, | 162, | 94, | 630, | 53 |
| 15 | B | 639, | 175, | 81, | 620, | 135 |
| 15 | E | 341, | 125, | 26, | 51, | 562 |
| 16 | B | 596, | 584, | 730, | 117, | 22 |
| 16 | E | 201, | 304, | 98, | 17, | 239 |
| 17 | B | 444, | 740, | 354, | 744, | 22 |
| 17 | E | 399, | 246, | 379, | 110, | 594 |
| 18 | B | 116, | 115, | 60, | 414, | 253 |
| 18 | E | 8, | 703, | 436, | 579, | 626 |
| 19 | B | 482, | 438, | 671, | 187, | 45 |
| 19 | E | 499, | 87, | 317, | 205, | 239 |
| 20 | B | 400, | 165, | 303, | 532, | 414 |
| 20 | E | 54, | 220, | 503, | 458, | 363 |
| 21 | B | 192, | 549, | 201, | 427, | 37 |
| 21 | E | 722, | 464, | 252, | 426, | 641 |
| 22 | B | 14, | 445, | 405, | 533, | 269 |
| 22 | E | 687, | 177, | 204, | 50, | 10 |
| 23 | B | 202, | 410, | 112, | 221, | 804 |
| 23 | E | 335, | 95, | 648, | 144, | 252 |
| 24 | B | 397, | 281, | 37, | 788, | 315 |
| 24 | E | 56, | 38, | 269, | 53, | 242 |
| 25 | B | 119, | 319, | 655, | 105, | 93 |
| 25 | E | 395, | 86, | 198, | 549, | 257 |
| 26 | B | 250, | 80, | 685, | 527, | 461 |
| 26 | E | 81, | 496, | 116, | 612, | 675 |
| 27 | B | 167, | 117, | 14, | 297, | 384 |
| 27 | E | 332, | 126, | 298, | 59, | 69 |
| 28 | B | 302, | 765, | 861, | 259, | 501 |
| 28 | E | 549, | 331, | 265, | 40, | 232 |
| 29 | B | 227, | 149, | 616, | 415, | 403 |
| 29 | E | 220, | 270, | 7, | 198, | 137 |
| 30 | B | 577, | 394, | 402, | 197, | 21 |
| 30 | E | 40, | 432, | 476, | 147, | 425 |
| 31 | B | 447, | 339, | 19, | 284, | 91 |
| 31 | E | 78, | 87, | 7, | 838, | 384 |
| 32 | B | 511, | 217, | 203, | 48, | 312 |
| 32 | E | 207, | 28, | 275, | 549, | 273 |
| 33 | B | 149, | 332, | 596, | 254, | 467 |
| 33 | E | 332, | 334, | 339, | 475, | 712 |
| 34 | B | 297, | 316, | 553, | 185, | 356 |
| 34 | E | 63, | 648, | 556, | 127, | 118 |
| 35 | B | 132, | 175, | 179, | 502, | 106 |
| 35 | E | 245, | 174, | 389, | 343, | 409 |
| 36 | B | 659, | 193, | 619, | 277, | 652 |
| 36 | E | 143, | 638, | 396, | 408, | 340 |
| 37 | B | 689, | 463, | 442, | 301, | 340 |
| 37 | E | 405, | 608, | 385, | 406, | 327 |
| 38 | B | 401, | 287, | 58, | 480, | 62 |
| 38 | E | 53, | 17, | 20, | 615, | 555 |
| 39 | B | 503, | 205, | 129, | 626, | 463 |
| 39 | E | 169, | 733, | 73, | 551, | 77 |
| 40 | B | 155, | 48, | 424, | 310, | 124 |
| 40 | E | 568, | 735, | 660, | 185, | 514 |
| 41 | B | 172, | 294, | 597, | 210, | 334 |
| 41 | E | 39, | 9, | 472, | 176, | 182 |
| 42 | B | 174, | 761, | 373, | 383, | 444 |
| 42 | E | 323, | 539, | 316, | 78, | 491 |
| 43 | B | 39, | 39, | 344, | 49, | 72 |
| 43 | E | 12, | 838, | 164, | 120, | 429 |
| 44 | B | 209, | 115, | 252, | 101, | 276 |
| 44 | E | 432, | 447, | 567, | 415, | 130 |
| 45 | B | 32, | 115, | 350, | 269, | 557 |
| 45 | E | 12, | 306, | 326, | 704, | 77 |
| 46 | B | 305, | 662, | 93, | 576, | 320 |
| 46 | E | 644, | 155, | 569, | 300, | 195 |
| 47 | B | 510, | 349, | 186, | 167, | 1 |
| 47 | E | 293, | 478, | 685, | 120, | 72 |
| 48 | B | 326, | 397, | 381, | 812, | 408 |
| 48 | E | 627, | 101, | 85, | 14, | 206 |
| 49 | B | 51, | 392, | 146, | 36, | 47 |
| 49 | E | 480, | 418, | 378, | 500, | 461 |
| 50 | B | 547, | 778, | 120, | 402, | 582 |
| 50 | E | 213, | 202, | 677, | 303, | 332 |
| 51 | B | 226, | 58, | 544, | 38, | 639 |
| 51 | E | 491, | 625, | 174, | 458, | 16 |

| DAY | Row | | | | | |
|---|---|---|---|---|---|---|
| 1 | C | 391, | 216, | 647, | 86, | 225 |
| 1 | F | 279, | 31, | 466, | 179, | 352 |
| 2 | C | 206, | 494, | 9, | 235, | 619 |
| 2 | F | 302, | 355, | 369, | 123, | 574 |
| 3 | C | 248, | 726, | 115, | 556, | 376 |
| 3 | F | 213, | 171, | 46, | 699, | 132 |
| 4 | C | 64, | 348, | 334, | 179, | 7 |
| 4 | F | 477, | 637, | 208, | 271, | 800 |
| 5 | C | 310, | 170, | 15, | 465, | 37 |
| 5 | F | 384, | 404, | 494, | 584, | 83 |
| 6 | C | 75, | 147, | 438, | 556, | 553 |
| 6 | F | 308, | 65, | 809, | 411, | 145 |
| 7 | C | 53, | 277, | 384, | 12, | 548 |
| 7 | F | 229, | 489, | 286, | 356, | 196 |
| 8 | C | 112, | 528, | 353, | 407, | 482 |
| 8 | F | 394, | 273, | 194, | 89, | 468 |
| 9 | C | 63, | 67, | 532, | 348, | 313 |
| 9 | F | 133, | 167, | 67, | 595, | 184 |
| 10 | C | 315, | 485, | 528, | 265, | 161 |
| 10 | F | 328, | 372, | 308, | 214, | 172 |
| 11 | C | 223, | 166, | 480, | 359, | 131 |
| 11 | F | 561, | 409, | 255, | 159, | 474 |
| 12 | C | 506, | 232, | 219, | 157, | 156 |
| 12 | F | 49, | 169, | 525, | 314, | 275 |
| 13 | C | 573, | 82, | 89, | 184, | 230 |
| 13 | F | 625, | 244, | 417, | 340, | 604 |
| 14 | C | 220, | 557, | 621, | 80, | 220 |
| 14 | F | 196, | 674, | 202, | 195, | 616 |
| 15 | C | 412, | 391, | 123, | 405, | 213 |
| 15 | F | 178, | 405, | 280, | 368, | 3 |
| 16 | C | 229, | 673, | 38, | 216, | 566 |
| 16 | F | 148, | 368, | 4, | 103, | 753 |
| 17 | C | 321, | 11, | 552, | 463, | 189 |
| 17 | F | 292, | 239, | 144, | 403, | 400 |
| 18 | C | 617, | 374, | 681, | 571, | 100 |
| 18 | F | 40, | 2, | 17, | 543, | 641 |
| 19 | C | 185, | 186, | 99, | 12, | 194 |
| 19 | F | 276, | 78, | 61, | 46, | 307 |
| 20 | C | 594, | 198, | 364, | 402, | 235 |
| 20 | F | 146, | 129, | 704, | 379, | 454 |
| 21 | C | 85, | 507, | 211, | 298, | 73 |
| 21 | F | 772, | 271, | 598, | 193, | 238 |
| 22 | C | 799, | 131, | 243, | 301, | 11 |
| 22 | F | 352, | 289, | 151, | 390, | 70 |
| 23 | C | 547, | 191, | 40, | 406, | 152 |
| 23 | F | 26, | 472, | 585, | 267, | 127 |
| 24 | C | 544, | 240, | 567, | 239, | 528 |
| 24 | F | 128, | 612, | 540, | 684, | 549 |
| 25 | C | 440, | 781, | 15, | 270, | 387 |
| 25 | F | 245, | 428, | 22, | 675, | 158 |
| 26 | C | 307, | 237, | 441, | 560, | 7 |
| 26 | F | 117, | 89, | 326, | 94, | 289 |
| 27 | C | 321, | 377, | 448, | 44, | 214 |
| 27 | F | 20, | 579, | 368, | 4, | 407 |
| 28 | C | 363, | 492, | 206, | 222, | 385 |
| 28 | F | 170, | 525, | 344, | 327, | 206 |
| 29 | C | 292, | 92, | 429, | 154, | 59 |
| 29 | F | 146, | 352, | 192, | 47, | 188 |
| 30 | C | 386, | 660, | 426, | 722, | 148 |
| 30 | F | 401, | 120, | 160, | 598, | 625 |
| 31 | C | 80, | 462, | 57, | 598, | 145 |
| 31 | F | 292, | 195, | 383, | 157, | 137 |
| 32 | C | 142, | 504, | 341, | 186, | 474 |
| 32 | F | 357, | 26, | 300, | 668, | 34 |
| 33 | C | 737, | 562, | 197, | 227, | 48 |
| 33 | F | 530, | 208, | 213, | 418, | 385 |
| 34 | C | 347, | 67, | 102, | 193, | 367 |
| 34 | F | 92, | 514, | 56, | 562, | 204 |
| 35 | C | 482, | 509, | 202, | 310, | 445 |
| 35 | F | 264, | 270, | 430, | 86, | 125 |
| 36 | C | 66, | 145, | 148, | 177, | 119 |
| 36 | F | 859, | 141, | 334, | 375, | 218 |
| 37 | C | 208, | 447, | 312, | 239, | 194 |
| 37 | F | 21, | 199, | 855, | 707, | 449 |
| 38 | C | 484, | 460, | 588, | 293, | 265 |
| 38 | F | 302, | 531, | 411, | 156, | 268 |
| 39 | C | 447, | 160, | 435, | 523, | 106 |
| 39 | F | 495, | 159, | 44, | 826, | 464 |
| 40 | C | 15, | 134, | 52, | 382, | 422 |
| 40 | F | 120, | 460, | 634, | 213, | 299 |
| 41 | C | 32, | 250, | 490, | 195, | 49 |
| 41 | F | 427, | 250, | 209, | 544, | 437 |
| 42 | C | 262, | 411, | 204, | 570, | 120 |
| 42 | F | 450, | 490, | 91, | 577, | 571 |
| 43 | C | 149, | 288, | 626, | 500, | 28 |
| 43 | F | 524, | 70, | 103, | 369, | 562 |
| 44 | C | 51, | 371, | 63, | 189, | 341 |
| 44 | F | 5, | 97, | 307, | 142, | 541 |
| 45 | C | 48, | 286, | 218, | 10, | 487 |
| 45 | F | 121, | 140, | 93, | 25, | 568 |
| 46 | C | 109, | 82, | 173, | 242, | 572 |
| 46 | F | 18, | 644, | 489, | 45, | 205 |
| 47 | C | 540, | 238, | 63, | 5, | 93 |
| 47 | F | 469, | 128, | 186, | 222, | 181 |
| 48 | C | 57, | 391, | 164, | 722, | 396 |
| 48 | F | 464, | 372, | 387, | 480, | 52 |
| 49 | C | 152, | 411, | 128, | 297, | 313 |
| 49 | F | 347, | 33, | 257, | 162, | 39 |
| 50 | C | 197, | 46, | 398, | 836, | 494 |
| 50 | F | 21, | 203, | 530, | 26, | 89 |
| 51 | C | 761, | 692, | 375, | 91, | 552 |
| 51 | F | 452, | 393, | 31, | 25, | 156 |

Answer Key Mapping:

| DAY | | | | | | |
|---|---|---|---|---|---|---|
| 1 | A | 600, | 33, | 410, | 188, | 252 |
| | D | 81, | 617, | 438, | 336, | 279 |
| | B | 233, | 418, | 81, | 262, | 341 |
| | E | 385, | 416, | 504, | 512, | 245 |

Day 1 Problems & Solutions

A

```
  721        581        851        919        666
- 121      - 548      - 441      - 731      - 414
-----      -----      -----      -----      -----
  600         33        410        188        252
```

Two rows (A–F) with five columns of solutions for each day.

**6600 Practice Problems**

# 100+ Days of Timed Tests

## Double Digit Addition & Subtraction

Grade 1-3

60

120 Pages

Ages 6-9

MATH DRILLS WITH & WITHOUT REGROUPING

**High Frequency Words**

# FRY's 1st 100 SIGHT WORDS

## PRACTICE BOOK

For Kindergarten & Grade 1

One

One

COLORING | Ages 4+ | TRACING

**Toddlers and Preschoolers Workbook**

# BIG LETTER TRACING

Hi!

Alphabets | Numbers

Ages 2-4

120 Pages

FINGER PRACTICE | COLORING

The 3 Minute

# GRATITUDE JOURNAL

Kind

# GRATITUDE ACTIVITY
## FOR KIDS

Thanks

Sight Words Tracing

# 100 MUST KNOW SIGHT WORDS

## PRACTICE BOOK For Kids

Aa Aa

can

I can do it.

Ages 3+

TRACING          COLORING

6600 Practice Problems

# DOUBLE DIGIT ADDITION

## 100+ Days of Timed Tests
### Grade 1-3

Math Drills

60

120 Pages

Ages 6-9

EXCELLENCE IN ADDITION          MATH DRILLS

6600 Practice Problems

# DOUBLE DIGIT SUBTRACTION

## 100+ Days of Timed Tests
### Grade 1-3

Math Drills

60

120 Pages

Ages 6-9

EXCELLENCE IN SUBTRACTION

Check out our other books!

## abcZbook Press

# Certificate of Excellence Award

# Triple Digit Addition & Subtraction
### (With Regrouping)

Congratulations!

You are a SuperStar!

By:

Date:

abcZbook